Bagriy & Co.

Alexei Tsvelik

Six Days

Reason as a Cosmic Phenomenon

Bagriy & Company
Chicago
2019

SIX DAYS:
Reason as a Cosmic Phenomenon
By Alexei Tsvelik

With
A Foreword by Alexey Burov
And
An Afterword by Michael Meerson

ISBN: 978-1-7337824-7-0 (Hardcover)
ISBN: 978-1-7337824-8-7 (Paperback)
ISBN: 978-1-7337824-9-4 (E-book)

Library of Congress Control Number: 2019953700

Sponsored by
Fr. George Florovsky
Memorial Research and Publication Center
At Christ the Savior Orthodox Church in NYC

Edited by Penelope Burt
Cover design by Natali Cohen
Book design by Yulia Tymoshenko

Published by
Bagriy & Company
Chicago, Illinois, USA
www.bagriycompany.com

Printed in the United States of America

CONTENTS

Foreword

Alexey Burov

Why is there something rather than nothing? Three centuries have passed since Leibniz asked this question. More than a few powerful minds have approached it, leaving it with or without an answer. Yet the question continues to entice, irritate, repel, alarm and arouse philosophical thought. Answering it engages the deepest parts of a person—thinking, believing, being grateful—as does any philosophical question. That does not mean the answer is arbitrary. On the contrary, it is, perhaps, fateful to the one answering, being determined on his basic qualities, by which the truth can be either illuminated or eclipsed.

Something differs from nothing in that it consists in a meaningful order. Chaos, the metaphysical chaos, is that which lacks order completely; therefore, it is nothing. Thus, one can rephrase Leibniz's question: *why order and not chaos?* Such a dichotomy implies that, to a certain degree, the world contains both, and that chaos by itself does not demand explanation, but that order does. What does demand explanation about chaos is that measure of it to which it manifests. In ethics, chaos is evil. Consequently, Leibniz's question can be rephrased a third way: *why is the world a mixture of good and evil?* Can the source of good be also the source of evil? And if not, then where does evil come from? Why did the source of good permit evil, if it is itself good? Leibniz offered his answers to this trifold question in his book *Theodicy* (1710), the only one published in his lifetime, but this is not the time nor place to go over them.

What are some possible answers to Leibniz's question? It's hard to say right off the bat, so we'll ask in a different way: what answers are impossible? Immediately, we can cross off the list the ones that take as a starting point some or other set of laws, even those still unknown to science. All the laws of this world are already grouped into the category of "something" and, thus, each one demands explanation themselves. No matter what the fundamental laws turn out to be, Leibniz's question places them into the category of "something existing" and asks—why are they in effect? Thus, we come to the fourth formulation of Leibniz's question: why are the laws of the world the way they are rather than some other way? Regardless of how many laws there are, known or unknown, by simple logic they are incapable of answering the question as to why it is they and not other mathematical or logical forms that partake in the structuring of the world.

All right, but if no law can serve as an answer to Leibniz, then what can? Could it be that the source of harmony in the world is chance, the totality of chaos? Could the question of cosmogenesis be resolved by framing it as *chaosogenesis*? By looking at the only universe known to us, having no data about any other possible worlds, could we accept or reject the hypothesis that at the foundation of Being lies a generator of worlds, randomly spitting out all thinkable and unthinkable universes with all possible laws, or even without any, and that the only reason we are discussing any questions in this and not another universe is because chance threw us here and not there? The author of these words considered this *chaosogenetic* version of the answer to Leibniz in an essay entitled "Genesis of a Pythagorean Universe" (2016), and came to the conclusion that the laws of the universe are too special: they are too simple, universal, precise, elegant, life- and thought-friendly, and highly conducive to discovery for us to accept blind chance as their explanation, even assuming an infinity of universes, churned out by a generator of some inexplicable origin. The attempt to explain the laws by chance is a complete failure. What choice of responses are we left with to the request to explain the laws? In fact, we are left with only two.

The first one is the answer that portrays the One All-Good Principle, the totality of Good, as the source and foundation of the world. Because the laws of the universe are, in this case, given by its Source, we can only think of Him as transcending the universe, as a transcendental God, akin to the way an artist transcends his painting, or the author his novel. It may even be that He created a multitude of universes, a multiverse, but this multiverse would be a completely different one than that of *chaosogenesis*. Each universe of the divine multiverse would reflect the perfection of the thought of its Author rather than meaningless chance. Further, the discoverability of laws can be comprehended as a realization of the Creator's desire to share the joy of this beautiful solution with us, sapient life forms that appear by His design and will in this universe. It is possible to see in the discoverability of laws a doorway to a certain union with God; thus Physics takes on a religious dimension, which played a key role in its birth and development. This answer to the question of Leibniz culminates in the totality of Good, which is the highest reality and the final singular foundation, the substance of all reality. In the Good there is nothing specific or partial, and therefore the question "Why is it this way, and not some other?" has no meaning in this context. There is nothing "other" for it beside the God except non-being. If one were to ask, in turn, what the existence of the Good is due to, another could only answer by pointing out how this is the final mystery, of which reason itself also partakes, it being the grandeur of reason.

There is one and only one self-consistent alternative to the above answer, the skeptic answer, an *a priori* mistrust and denial that reason could even hope to approach so lofty a question. From the point of view of the skeptic, reason can only be relied on within the limits of sensory experience and the natural science tied to it. The question of Leibniz, however, in all its formulations, takes us outside the "legitimate" sphere of reason to the place where there are no grounds to trust it. The theistic answer breaks down under the skeptic's *a priori* mistrust of thought at this altitude. In support of his mistrust, the skeptic can present quantum physics with its paradoxes, which call into question even in such truths as the rule of the excluded middle. The electron turned out to be both a point and spread out over space, for instance. "Yes," says the skeptic, "reason was able to describe quantum objects, but only at the cost of rejecting a whole array of its truths, which seemed unshakable until then. However, this success only became possible due to experiments that refuted some hypotheses while supporting others. The experimental data forced the theorists to reject certain timeless truths and to feel out such theoretical schemes that without this data nobody would even care to discuss. The habitual truths of reason already fail at the level of the electron, and you are trying to conquer the heavens with them, and without experiments that could correct you, to boot."

The skeptic position may seem invulnerable, but it does have a serious flaw: skepsis is ethically impotent; it lacks the mustard seed of faith, without which the mountains of ignorance cannot be moved. In denying meaning to the question of the nature of laws, skeptics, whether they wish it or not, deny the high and independent value of the pursuit of the cognition of these laws. The only values of fundamental science that a skeptic could support, without surrendering skepticism, pertain to its technological applications and its role as a particular kind of entertainment or sport. These values, however, are foreign to science itself, to the selfless service of its truth. It is not surprising that among the founders of physics, there have been no skeptics at all, only theists and deists. As fundamental science develops, the forefront of its activity moves farther and farther from the scope of application, and the argument for its practical usefulness loses its power: it is hard to imagine that any practical usefulness can emerge from a better understanding of the Higgs boson or the evolution of the early universe, not to mention anything at the level of string theory or the Planck scale; no inspiration can be drawn here from skepsis. The only role that a consistent skepticism leaves to thought is that of serving life, rather than ruling over it. Sovereignty can only exist in the name of something noble, and the skeptic has nothing nobler than the life. But life ends with death; skeptics are sure of that.

And if this is so, then life is doomed to a meaningless end. The skeptic is playing a game, which he knows he will lose. All his efforts are in vain. Any inspiration that goes beyond life is incompatible with skepsis.

It is a common belief that skepticism is the foundation of science, that science demands putting everything to the test of doubt, that such a category as faith is foreign to it. Above, I tried to show that this common opinion is a mistake. Of course, science demands the rigorous and multi-faceted testing of theories and hypotheses, which it does not take on faith. But that does not mean that it is a stranger to faith. Without faith in truth as the highest value and the great potency of reason, science would have always remained on a purely empirical level without so much as an inkling of all those horizons that would be waiting to be by opened up by theoretical thought. Skeptics have never been absent or far away. Had Kepler, Galileo, Descartes, or Newton heeded them, humanity would be in the same place today as it was in the sixteenth century, if not below that level.

The fantastic success of mathematical physics should have been enough to serve as a powerful proof for the theistic rationalism of its founders, to the chagrin of the skeptics. That's why it is so odd that during the last half century, when scientific discoveries and the technical inventions based upon them overflowed like a cornucopia, quite the opposite happened: skeptics almost entirely displaced theists from the scientific Mount Olympus. Those who call themselves atheists, agnostics, materialists, and naturalists, as well as many of those who are described with the word scientism, are mostly just the varieties of skeptics. How did this happen, and does not this loss of roots forebode a degradation of physics? These are important questions which I will put aside, to conclude instead with a few words specifically about this book.

Its author, Alexei Tsvelik, like the author of these words, belongs to a dwindling tribe of rationalist theists among physicists, who share, in general, that synthesis of worldviews—the Platonic and the Biblical—that was so characteristic of the founders of physics. We share this synthesis not because we were nurtured in it: on the contrary, we both grew up in the USSR, where any deviation from state atheism, even mere conversations on the topic of religion, meant repression. At home and in school we were presented with atheism as the mandated worldview, as something absolutely obvious. So the nurturing we received was 100% militant godlessness. We were brought to theism by reading the literature forbidden by the communist state, by having conversations with the wise people we were lucky enough to meet, and by our own deliberations, all when we were already adults in our twenties. The power that brought us to theism was the same that brought us to physics: the thirst for the truth and our belief in our abilities along that path. We pursued a worldview with no lesser tenacity than we pursued physics, by means of lengthy meditations over evidence of very different sorts, including the scientific. The book of my friend, Alexei Tsvelik, a leader of the theoretical group of the Brookhaven National Laboratory (USA) and laureate of the Alexander von Humboldt prize, unpacks a part of his theistic arguments, the part that starts from natural science, physics, and biology. Written in the clear language of a theoretical physicist, it presents in itself a thoughtful meditation on life as an entity on a cosmic scale, interesting both on its own merit and for the metaphysical data it offers.

This is the task that Alexei Tsvelik long ago set before himself, for which he long ago set off in pursuit of arguments, both pro and contra. In his own words:

> *We have to find out whether our arrival in this world was just an accident or the result of a purposeful process.*

This pursuit has turned out to be against the party line, in the West as well as in the USSR. Therefore this book can be thought of as anything but following any mandated or indoctrinated norms. This book is unlikely to prove of interest to ideologues, meaning those who pursue ready-made, correct ideas, who pay no mind to the quality of thought, since the book is a product of different priorities. I think that it can be interesting only to those readers who in regard to the most important things are akin to its author—those for whom truth follows no party line, those who find joy in the power of an argument and the testing of this power against the most powerful of objections. It is this kind of reader that I wish for this book.

Happy is the man who can recognize in the work of Today a connected portion of the work of life, and an embodiment of the work of Eternity.

—J. C. Maxwell

INTRODUCTION

To the present day observer the position occupied by the human race in the grand design of things appears as miniscule. Many believe that science just reinforces what already appears to be obvious. Does not it tell us that even in the history of our planet, which by itself constitutes but a small speck in the vast Universe, the history of our species occupies just a tiny part. However, since the subject is so important it is worth looking into the evidence more carefully.

To achieve a proper understanding of our origins and to appreciate the position we occupy, one needs to consider life and intelligence as cosmic phenomena, placing them into the wider context of the history of the Universe. To do this one needs to bring together facts from the many scientific disciplines such as physics, chemistry, geology and biology which together constitute the physical and natural sciences. As I will try to demonstrate, a careful analysis will rather reinforce what may be called the anthropocentric view.

In my narrative I only use results about which the scientific community has reached a consensus; the reader will not find here any sensational material or any theories idiosyncratic to the author.

Since I am going to rely on science in this narrative, it is proper to start the discussion from science itself. No train of thought or discussion ever starts from nothing; we always make initial assumptions. To avoid confusion, it is better to lay them bare at the very beginning. The natural sciences are no exception: they also have their own creed. It is not always clearly and openly presented when scientists talk about their discoveries or give their interpretation of what is going on in the world. But science does assume that world events occur according to rules which scientists call the Laws of Nature. These rules are thought to be independent of our will and the aim of science is to obtain a knowledge of these laws. Knowledge of the laws of nature enables us to predict future events—and also to reconstruct the events of the past. The subsequent narrative will rely heavily on such reconstructions.

It is important to bear in mind that science does not just systematize facts about the world, bringing them into some tidy order. The system and the order are considered to be good only insofar as they allow us to make accurate predictions. One has to remember that according to the modern view these laws are not deterministic but statistical, thereby admitting a certain degree of flexibility. There is an intricate play in nature between chance and necessity, which together make our world so rich.

Reliable knowledge is predictive and verifiable. Naturally, by insisting on reliability, we narrow the field of science. Hence, one cannot claim that the natural sciences are able to encompass every aspect of the truth, especially when it concerns the affairs of mankind.

Belief in natural laws underlies everything that natural scientists do. Even when they become equivocal about this, their actions speak louder than their words. However, belief is the union of evidence and hope, and no amount of evidence will persuade every skeptic.

There is a subtle question as to whether the term "world events" encompasses everything, including human behavior, or whether there is a line (or even lines) separating the world of inanimate matter from creatures with intentions and emotions and, eventually, from those, like us, with intelligence and the capacity for reflective thinking. One wide spread position called physicalism or naturalism maintains that there is no essential difference between the human sphere and the rest of the world and hence the natural sciences can (at least in principle) explain it all. On the other hand, there are those who think that science is just a human invention, making pronouncements about human nature that are obviously wrong, and on this basis they do not take it seriously even when it speaks about inanimate objects.

I belong to neither of those camps. In this Introduction it will be enough for now to say that I believe that the natural sciences have complete authority when it comes to inanimate objects. At the same time, I think that this authority becomes shaky when we are dealing with living creatures and shakier still when we deal with those aspects of human life which are special to us as humans, such as history and economics. Our ultimate purpose, however, will be to range more widely.

There is a widespread belief that the principal adversary of science is religion, with its presumed belief in the "supernatural." In fact, the concept of the supernatural is not a necessary attribute of religion, as may be ascertained from the fact that the foundations of science were established by deeply religious people. The very idea of natural laws—without which science would not exist—was conceived in the religious society centered on the ancient Greek philosopher and sage Pythagoras. These ideas were later developed by Plato and Aristotle, both religious people, although they did not adhere to any popular religion of the day. The founder of modern European science Isaac Newton was also a deeply religious man, as were such other great scientists as Michael Faraday, James Clerk Maxwell, Max Planck, Isidor Rabi, Georg Cantor, Kurt Gödel, and Werner Heisenberg. And although Albert Einstein, Erwin Schrödinger, Eugene Wigner, and Paul Dirac were not religious in the conventional sense, it would be preposterous to label them as atheists or even agnostics.

Perhaps the following quotation from Albert Einstein would serve as a good illustration of this thesis.

> The interpretation of religion, as here advanced, implies a dependence of science on the religious attitude, a relation which, in our predominantly materialistic age, is only too easily overlooked. While it is true that scientific results are entirely independent from religious or moral considerations, those individuals to whom we owe the great creative achievements of science were all of them imbued with the truly religious conviction that this universe of ours is something perfect and susceptible to the rational striving for knowledge. If this conviction had not been a strongly emotional one and if those searching for knowledge had not been inspired by Spinoza's *Amor Dei Intellectualis*, they

would hardly have been capable of that untiring devotion which alone enables man to attain his greatest achievements (cited in Jammer 2002, 117)

The ontological status of natural laws. One of the reasons for the apparent conflict between science and religion is a misunderstanding of the ontological status of natural laws. Experience acquired in the pursuit of the natural sciences suggests that the laws constitute a logical structure for constantly changing world events. In other words, scientists believe that the world changes, but the laws do not. If it were otherwise, science would have no predictive power. To maintain that the Universe possesses a logical structure is tantamount to maintaining that it is ruled by Reason or Logos.

The following statements, the first by Albert Einstein and the second by the famous British philosopher and mathematician Alfred North Whitehead, illustrate our point.

> ...ultimately the belief in the existence of fundamental all-embracing laws also rests on a sort of faith. All the same, this faith has been largely justified by the success of science. On the other hand, however, everyone who is seriously engaged in the pursuit of science becomes convinced that the laws of nature manifest the existence of a spirit vastly superior to that of men, and one in the face of which we with our modest powers must feel humble (Einstein to Phyllis Wright, 24 January 1936, quoted in Jammer 2002, 93).

> In the first place, there can be no living science unless there is a widespread instinctive conviction in the existence of an *Order Of Things,* and, in particular, of an *Order Of Nature...*

...the inexpugnable belief that every detailed occurrence can be correlated with its antecedents in a perfectly definite manner ... must come from the medieval insistence on the rationality of God...

...My explanation is that the faith in the possibility of science, generated antecedently to the develop-ment of modern scientific theory, is an unconscious derivative from medieval theology (Whitehead 1967 [1925], 3–4, 12–13).

This governing logical structure is literally not of this world, by virtue of having a status different from that of the events. We deduce this structure from the obser-vation of the phenomena—not seeing it with our eyes or hearing it with our ears, but deducing it through our intelligence by means of hypothesis and analysis. I say that the Laws of Nature are not of this world, but con-stitute their own realm, because they are not located at any particular point in time or space. They are not things or events, but they direct events and therefore one may speak with confidence about their preexistence and independence of the material content of the Universe. The proper analogy for the relationship of the natural laws to matter would be that of the blueprint to the building or software to hardware.

The reader may be surprised by such an idealistic view of science, but it is implicitly contained in every physics textbook. According to this view space and time together with their material content do not govern themselves; their behavior and fate is determined by a system of laws which are themselves atemporal and all encompassing. In other words, the laws are universal. This idea is at the foundation of modern physics and has far-reaching consequences.

The laws are "out there" in the sense that they are not our invention, nor are they social constructs, although our knowledge of them is necessarily limited and changes with time. No self-respecting scientist would say that he or she has invented some natural law; laws are not invented, but discovered, just as Columbus discovered America for Europeans, although it had been there all the time, long before his voyage. Just as Columbus mistook America for India because his theory was wrong, we can also be confused about the real meaning of our discoveries; but our understanding is improved through a continuing process of criticism, verification, and argument.

One may ask: can we ever feel confident of any knowledge? Not only members of the general public, but some philosophers believe that this is not possible, that every new scientific epoch cancels the achievements of the previous one, since "paradigm shifts" in scientific thinking allegedly create impenetrable barriers in our intellectual development.[1] If this were true, science would have no say in such matters as the ones discussed on these pages. I think, however, that this point of view is based on a misunderstanding. New developments in science do not cancel the achievements of the past, but rather put limits on their validity and accuracy. The Theory of Relativity, for instance, has proven the Newtonian postulate of absolute time to be wrong, but in doing so it has not invalidated all of Newton's achievements and results. In a similar fashion quantum mechanics disproved the absolute determinism of classical mechanics. Notwithstanding this, anyone interested in the motion of macroscopic bodies whose speed is much less than the speed of light can still rely on the laws of mechanics formulated by Newton more than 300 years ago. In fact, that is exactly what is done by engineers and meteorologists.

The same goes for chemistry: even though we do not yet understand how to unite quantum mechanics with gravity, it is of little concern to any practicing chemist and, I dare say, will remain so even after this unification has been achieved.

This book assumes that there are areas of our experience where our knowledge is reliable. And I will try to show that even this limited knowledge allows us to draw far-reaching metaphysical conclusions about the overall comprehensibility of the universe.

Evidence for the atemporal and universal nature of the laws. Let us now come back to the idea of universality. We have ample evidence for believing that the laws of nature as we know them have remained the same throughout the history of the Universe and are the same in all the regions that we can observe. This idea has not been accepted without challenge. The evidence in its favor comes, however, from the spectroscopic analysis of remote cosmic objects. Atoms of various chemical elements emit or absorb electromagnetic radiation (this includes radio waves, infrared radiation, visible light, and X-rays) at a particular set of frequencies. Each element—and, in fact, each molecule—displays a unique pattern of emitted radiation, constituting a kind of "fingerprint." When astronomers analyze light from remote stellar objects they find the same spectral patterns as here on Earth. No matter how far away these objects may be—light years, thousands of light years or even billions of light years away—we still see the same patterns (shifted as a whole to the red side of the spectrum). This kind of analysis informs us about the chemical content of these remote objects and also suggests that the laws responsible for the composition of the elements are the same throughout the observable Universe. Since light travels with a finite speed, by looking farther away we look deeper into the past. This means that by observing the same spectral patterns in objects that are billions of light years away we can ascertain that the laws of physics have remained unchanged over all this time.

A thought experiment illustrating the atemporality of the laws. Since the concept of natural laws is at the center of this discussion, I consider it crucial that my readers get a proper grasp of it. Hence I invite them to conduct a thought experiment:

Let us imagine a very young Universe, at a time shortly after the Big Bang, when stars and galaxies had not yet been formed. There are not even any complex atoms, only elementary particles. Imagine that some higher power placed some incorporeal spirit in this young Universe and gave it our physics textbooks. Obviously the knowledge of physical laws contained in these books is rather incomplete. But even so, by reading them our angel would be able to extract enough information to conceive a broad outline of future developments. The angel would be able to predict that the expanding Universe will cool down and that matter will organize itself into ever more complex forms: elementary particles will form atoms of hydrogen and helium and the atoms will form dense clouds which will become stars. Matter that remains outside the stars will continue to cool down but the temperature in the stars will increase, giving rise to thermonuclear fusion. The fusion of hydrogen nuclei will create the nuclei of heavier elements, and so on. Just by reading those physics books the clever spirit would be able to make these predictions *well before the actual events took place.*

This thought experiment demonstrates that physical laws are not inherent in material objects, since they predict the formation of these very objects. And I hope it is clear now why I believe that the laws that govern the events and the events they govern are fundamentally different entities. This brings us to the main thesis of my narrative: idea precedes its material incarnation, and since in its unity it contains all stages of the entire process of incarnation, from the materially simplest to the increasingly more complex ones, we are justified in saying that the material complexity grows from material simplicity only due to the rich intelligence encoded in the program.

The science *de facto* operates on the assumption that material world, the world of things we can see, hear, taste, and touch, is not the only reality. There is also a reality that opens itself only to our intelligence. This certainly speaks against materialism, but does not necessarily speak for a belief in God. One can argue—and this argument has been put forth before—that the laws of nature constitute an impersonal force, lacking any awareness or concern for our existence. If one can call this god, it would be the indifferent god of Spinoza, not the personal God of Abraham, Isaac, and Jacob. For a religious person the crucial question is whether this Platonic world of ideas supplants God or belongs to God as an attribute or essential to God's nature.

To answer this question we need to establish what the relationship between natural laws and humans is. We need to find out whether our arrival in this world was an accident or the result of a purposeful process. We need to study carefully our relationship to this world, and especially our ability to extract reliable knowledge of it. So I invite the reader to follow me on a journey to discover how we can go about answering this question.

DAY ONE:

Formation of space-time, emergence of the four fundamental forces. Self-organization and self-destruction as the leading principles of cosmogenesis. Separation of light from darkness. Birth of galaxies and stars. Stars as alchemical furnaces

When the gaze of our reason tries to penetrate into the past it encounters a boundary. There are serious reasons to believe that what we call the Universe—that is, everything that we can somehow perceive and know—had a beginning. The majority of scientists refused to take this idea seriously as little as 60 years ago, and there are still those who dispute it now. Whether our Universe is unique or is one of many, whether our time had a beginning or we just inhabit one aeon preceded by others is a matter of debate. Whatever the truth about the ultimate origins of our world, one thing is clear: there was a time when the Universe was very simple and all its stuff was simple, without structure and unimaginably dense and hot. Whether this was the very beginning of time, or there had been anything before, is not clear, since our theories become questionable at times of the order of 10^{-43} seconds (the Planck time) from the presumed Beginning. We are more certain about the Universe's subsequent history, which cosmologists see as one of expansion and increasing complexity: the Universe goes from structureless and hence materially simple to highly structured and therefore materially complex.

At this point I should remind the reader of some basic ideas about space and time and their relation to matter. The area of physics dealing with these notions is called the General Theory of Relativity (GTR). Its foundations were laid down by Einstein at the beginning of the previous century and it has subsequently been developed by many brilliant minds, including Alexander Friedmann, George Gamow, Kurt Gödel, Robert Oppenheimer, Stephen Hawking, and Roger Penrose. According to GTR, time, space and matter are inextricably connected to each other. This was a revolutionary idea; before Einstein time and space had been considered either as independent absolutes (Aristotle) or as forms of human perception (Kant). The relativity of time manifests itself in the fact that its pace is different for observers moving with different speeds with respect to each other. Sometimes these ideas are difficult to digest for those who are used to thinking about time as some kind of universal register of passing events indifferent to their content. Careful analysis reveals, however, that this perception of time is oversimplified. Time is indeed relative and time intervals between events depend on the circumstances of the observer. One 24-hour day of an astronaut in a spaceship moving with a speed of 9/10 of the speed of light with respect to the Earth would be perceived by us as 55 hours. So if we were able to see him, he would appear to us to be moving at a very leisurely pace inside his ship. When astrophysicists say that the current estimate for the age of the Universe is 13.4 billion years, they use as the reference point our Earth as it is now. (For a light wave emitted in the Beginning, the time passed since then is exactly zero…)

The dependence of time on the circumstances (the reference frame) of the observer makes it necessary to consider space and time together as a *space-time continuum*. This continuum is four-dimensional[2] and its metric properties are described by a four-dimensional geometry. The geometry of space-time is inextricably connected to its material content; in other words it depends on the stuff that fills it. One of the effects of this connection is that the pace of time in the vicinity of gravitating bodies slows down (Fig. 1). The denser the matter the slower is the pace of time. Hence when we go back in time to the initial state of infinite density, time slows down and comes to a complete halt (at least if one does not take into account the effects of quantum mechanics which we still do not know how to do in a way that is consistent with GTR).

Many people are baffled by the suggestion that there can be different geometries. How can one imagine, for instance, that parallel lines, which are defined as non-intersecting, do indeed intersect? However, these things are not as complicated as they at first appear if we realize that we encounter different geometries even in our everyday experience. Compare, for example, the geometry of a globe with the geometry of a sheet of paper with its plane surface. The fact that they are different does not surprise us, and yet the surface of the globe illustrates one of the deepest concepts of non-Euclidean geometry. It provides an example of what is called a closed space without a boundary. Indeed, an ant moving on this surface will never encounter any boundary, although the space over which it travels is finite (closed).

As for parallel lines, the pattern of "parallels" on the globe provides us with an image of what is meant when

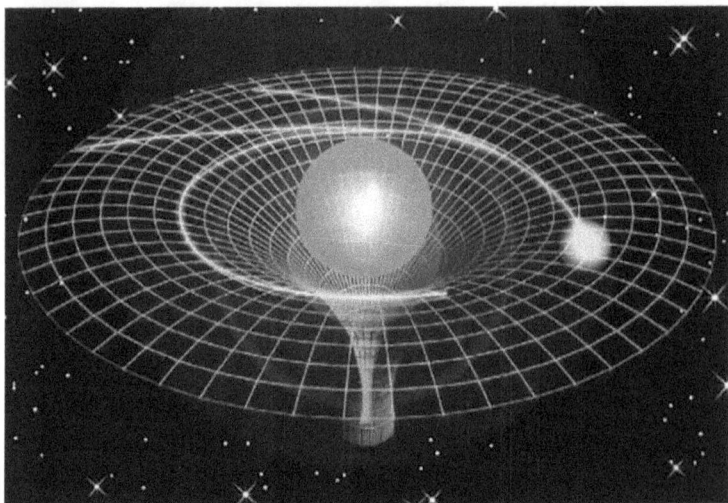

Fig. 1. A cartoon illustrating the change
of geometry near gravitating bodies.

mathematicians say that parallel lines in curved space may intersect. There are two types of parallel lines: lines with fixed latitude (parallels) and fixed longitude (meridians). They are perpendicular to each other. The meridians meet at the poles where the distance between them shrinks to zero.

A spherical surface provides an example of a *curved* space with only two-dimensions. Our space-time has four dimensions, which makes things more complicated. An additional complication comes from the fact that although time is linked to space, it still remains in some sense special. So it is no wonder that it is more difficult to imagine. But the human mind can work even with things the imagination fails to picture. The experience of mathematicians and physicists bears this out.

After this brief discussion of the basic concepts of time and space we may return to the main discussion. As I have said, some questions about the Beginning are not

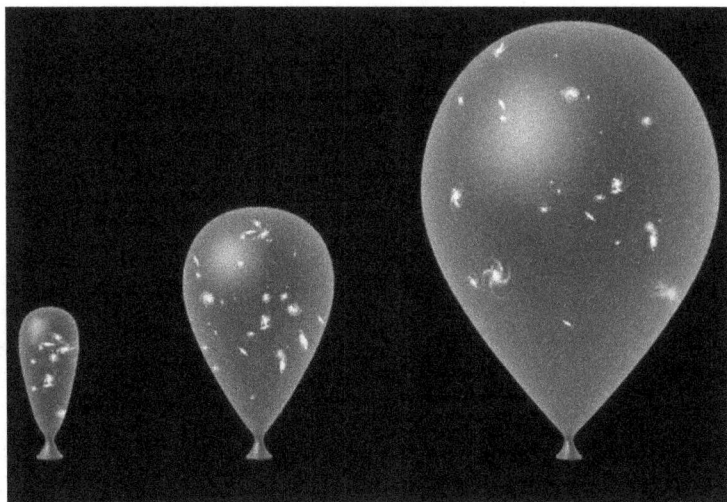

Fig. 2. Expanding space.

settled and may never be. This problem notwithstanding, the notion of a Beginning is not devoid of meaning. To those who ask what happened *before* the beginning of time I remind them that the notion of "before" refers to time and cannot be used if time did not (yet) exist. In other words the question "what happened before the beginning of time?" simply does not make sense (see St Augustine in Further Quotations to Day One).

Astronomical observations suggest that our space is perpetually expanding in such a way that the distances between all objects grow with time and the further away objects are from each other the faster they recede from one another. A graphic image of this is given by pictures drawn on the surface of an inflating balloon (Fig. 2).

The evidence for this expansion comes from the so-called "red shift"—the fact that the entire pattern of spectral lines from remote objects is shifted to the red side of the spectrum (the Doppler effect). These observations are in complete agreement with the predictions of GTR. As was first established by the Russian physicist Alexander Friedmann at the beginning of the 1920s, the equations of GTR describing the entire Universe do not allow for a stable solution. These equations state that the Universe began its run from a point-like state where all its mass was contained in a space of zero volume (the singularity). As I have mentioned above, this has been disputed on the grounds that GTR in its present form does not take into account quantum effects which are thought to become important close to the singularity. At present, the theory is considered as incomplete since it fails to provide a consistent description of the Universe as it was at a miniscule portion of a second from the Beginning. However, this incompleteness is not important for our story; it suffices for us to know (and here all our theories agree) that at about one-millionth of a second after the Beginning the newborn Universe did not contain any complex structures, only elementary particles. So it was structureless, "without form and void," full of matter in an unimaginably dense and hot state with particles moving practically at the speed of light. In this primordial hot soup of particles no stable complex structure could exist; everything was in a state of flux and change.

The observational evidence for the hot Universe is provided by the so-called *relic radiation* (Cosmic Microwave Background or CMB) discovered in 1964 by American physicists Arno Penzias and Robert Wilson at Bell Laboratories. This observation gives us a snapshot of the Universe at 380,000 years old, just at the moment when "light separated from darkness," that is, when light and matter (protons and electrons) practically ceased to interact (Fig. 3).

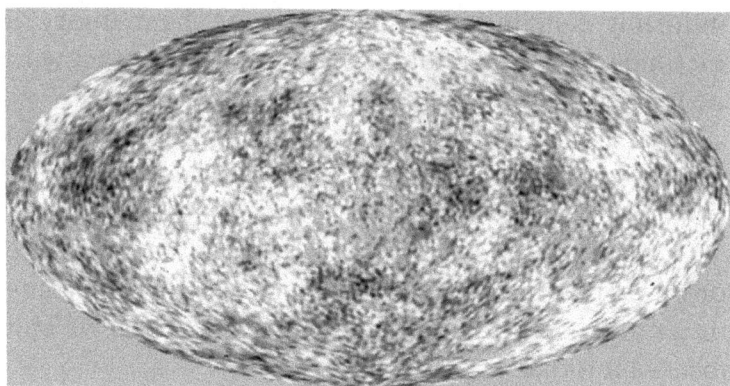

Fig. 3. Photographic image of the CMB map.
The variations of the temperature are greatly exaggerated.

The CMB is interpreted as radiation that has reached us from a time well before the formation of stars and planets, when the universe was filled with dense and hot ionized plasma, containing free protons and electrons. Since electrically charged particles not only emit, but also absorb light, the universe was not transparent for light; it was like a dense hot fog. This fog was in thermal equilibrium with itself, that is, the temperature was more or less the same everywhere[3] and—what is more important—the temperature of the radiation and the matter was the same. However, due to continuing expansion the universe eventually cooled down to a temperature where protons were able to capture electrons to form electrically neutral hydrogen atoms. The plasma became an electrically neutral gas. The energy of photons was no longer sufficient to ionize the atoms and the gas became transparent. As a result, the primordial light decoupled from the "darkness," that is from the hydrogen matter. The remaining photons were left to propagate freely, and their temperature has been decreasing ever since, but the radiation remained uniformly distributed in space keeping the memory of the times when the photons were in close contact with the matter.

The "separation of light from darkness," or photon decoupling, marks another important stage in the development of the Universe. From that time on one can no longer speak about a common temperature for all matter in the Universe. Some stuff will become hotter and some will continue to cool down as we still observe it today.

It should be noted that although the hot Universe was simple and devoid of any stable structures *it was not chaotic* (this point is discussed at length by Roger Penrose in his book *The Road to Reality* [2005, ch. 27]). It was in a state of low entropy,[4] a state of incredible potential full of unrealized possibilities. With the passage of time the potentiality will become reality *through a process of self-organization directed by the Law*.[5]

Here is an illuminating quotation from John Barrow and Frank Tipler's classic, *The Anthropic Cosmological Principle*:

> Twentieth century physics has discovered that there exist invariant properties of the natural world and its elementary components which render inevitable the gross size and structure of almost all its composite objects. The sizes of bodies like stars, planets and even people are neither random nor the *result of any progressive selection process* [the italics are mine—A.Tsv.], but simply manifestations of the different strengths of the various forces of Nature. They are examples of possible equilibrium states between competing forces of attraction and repulsion. (Barrow and Tipler 1986, 322)

By the time of the "separation of light from darkness" the Universe has become more structured. By that time protons have combined with electrons to form hydrogen atoms, hence establishing the basis for the subsequent formation of all chemical elements. This is one of the first stages of an ongoing process of the *self-organization* of matter into systems of ever increasing complexity. This universal process goes on side by side with the opposite process of destruction and degradation. Like Eros and Thanatos, these two principles are the prime movers of everything that happens in our world.

The process of self-organization is conditional in the sense that its existence requires a very special arrangement of natural laws. In particular, it is required that forces acting on very different space-time and energy scales work in concert. This requires an incredibly precise balance whose possible origins we will discuss later on. But it would not be an exaggeration to say that every step of it looks like a miracle. Whenever simple parts combine into something more complex, be it the formation of atoms of the elements from protons, neutrons, and electrons, or the formation of complex molecules from atoms, or the formation of living cells from complex molecules, you will find that its conditions of possibility are based on some very special structure of the natural laws.

Destruction and degradation on the other hand do not require any special arrangements. The scientific name for these processes is entropy growth, the growth of disorder. The second law of thermodynamics states that any physical process is accompanied by the loss of a system's ability to perform useful work—a sort of loss of creative power. Although energy does not disappear, it gradually becomes worthless. It may appear to be a paradox, but the newborn Universe, being simple and without structure, contained more order (less entropy) in its simplicity than the world of the present day which is so full of forms and structures. All this complexity occupies just a part of our world; it has emerged at a price, and we have been paying for it by the irrevocable degradation of another part. That is what the second law of thermodynamics states and it is a powerful argument against the eternity of the Universe.[6]

How does the self-organization proceed? First, one has to emphasize that this process has nothing to do with Darwinian evolution and its mechanisms of natural selection and random mutation When two atoms of oxygen create a molecule the outcome of the process is not selected from a random set; it is predetermined. Darwinian evolution can act only on living organisms, since it requires those who are unfit to die. Hence it cannot act on inanimate matter. As I will argue, the long process of emergence of ever more complex systems preceding the appearance of life owes itself not to chance, but to the special properties of the natural laws.[7]

Self-organization is a product of the action of different physical forces. Physics distinguishes four different fundamental forces or interactions: the gravitational, the electromagnetic, and the strong and the weak intranuclear forces.

The gravitational force acts between all material bodies; even those bodies which, like photons, do not have mass are subject to gravity, which bends the path of light. Although it is the weakest of all the four, its action decreases slowly with distance and therefore it is responsible for the global architecture of the Universe— the arrangement of the galaxies and stars and the orbits of cosmic bodies such as planets and comets.

The electromagnetic force is stronger than gravity, but its range of action is restricted by the fact that the total electric charge of macroscopic bodies is usually zero (the positive charges of protons are balanced by the negative ones of electrons). The electric force between elementary charges, such as a proton and an electron in a hydrogen atom, is many orders of magnitude stronger than the gravitational one, but the electric force between two cosmic bodies whose net electric charge is close to zero is miniscule in comparison with their mutual gravitational pull.

The electromagnetic force is responsible for all chemistry, including the chemistry of our bodies. Phenomena such as visible light, radio and micro waves, infrared and ultraviolet radiation, X- and gamma rays are all of an electromagnetic nature as well.

The weak and strong nuclear forces act on distances of the order of the size of atomic nuclei. Even the electrons in atoms do not, for the most part, experience their action; they are kept bonded to the nuclei by the electromagnetic force. The weak and strong interactions are responsible for the formation and stability of atomic nuclei and for their transformations, including nuclear fission and fusion. These processes are accompanied by strong releases of energy and are the source of stellar radiation. The reactions of thermonuclear fusion are responsible for the formation of elements heavier than helium and hydrogen.

Combinations of fundamental interactions create secondary forces. The magnetic exchange interactions, covalent and hydrogen bonding, and the van der Waals force are all offspring of the electromagnetic force and quantum mechanics.

It is believed that in the earliest stages of the formation of the Universe the interactions which look so different now were virtually indistinguishable. Their individuality has emerged with the decreasing temperature of the Universe through a series of phase transitions. A phase transition is a transformation in which a physical body changes its internal structure without changing its material content. The most familiar example of such a transition is when water freezes and turns from a liquid into a crystal (ice). The phase transitions that shaped the fundamental interactions into their present form, however, were of a more exotic kind; they resembled the transition into a superconducting state that some metals exhibit whereby they become perfect conductors capable of conducting electricity without any loss. The electromagnetic force inside a superconductor behaves differently than it does outside of it, so that the interaction between magnets, usually a long-range interaction, becomes a short-range one. Likewise, the weak interaction becomes a short-range force after one of such phase transitions. We can say that we live inside an enormous "weak-force superconductor," which encompasses the entire Universe.

We physicists have a special name for the process of individualization described above: we call it "symmetry breaking." In the earliest stages of the Universe the four interactions were indistinguishable, so they could be interchanged and it would make no difference. This is a state with high symmetry. When the temperature in the Universe dropped below a certain value the symmetry between, say, gravity and the other three forces got broken. In this broken symmetry state, gravity then emerged as a special force, different from the others. Then another transition occurred and the strong interaction split from the electroweak one and at last the weak and the electromagnetic forces were differentiated. As a result of these separations, each force acquired its own domain where its action was predominant. Physicists characterize these domains by the amount of energy exchanged in the elementary processes that involve this or that interaction. Stellar and planetary bodies act on each other predominantly by gravity; the properties, dynamics, and behavior of liquids and solid bodies are directed by electromagnetic interactions; the shining of the Sun and the stars is due to the nuclear forces. This process of self-organization requires that all the forces act harmoniously both *with respect to each other and within each domain*. The first is achieved by what is now called fine tuning; the second, as I will explain at length later, is achieved by the very special construction of the physical laws.

To get a foretaste of what this all means, let us look at one of the simplest acts of self-organization—the formation of hydrogen atoms. They consist of one positively charged heavy particle—a proton and one light negatively charged particle—an electron, which is bound to the proton by the electric (Coulomb) force. Hydrogen is the first and simplest element in the periodic table of elements. It plays an enormous role in chemistry and biology and also constitutes the first step in the formation of heavier elements. As I have stated before, stable hydrogen atoms emerged approximately 380,000 years after the Big Bang when the temperature dropped to such a degree that the electric force was capable of holding atoms together. Could we imagine a world where even such simple structures as hydrogen atoms would not form? Easily. For instance, no atoms would form in a world of one time and four spatial dimensions or higher because the electric force would not create a bound state of charges of opposite sign. On the other hand, in a world of one time and two spatial dimensions this force would be too strong to allow the atoms to ionize. This would also be a problem for chemistry because chemical reactions proceed via the exchange of electrons. So, our world has the right number of spatial dimensions—three, which allows atoms to be both stable and unstable depending on the circumstances.

The Birth of the Periodic Table of the Elements

The "separation of light from darkness" left the Universe in a state where it was filled with clouds of hydrogen and helium gas. Such a state is unstable: gravity tends to amplify any inhomogeneity, so that in places where the gas has a higher density it becomes ever denser and in places where the density is lower it further decreases. So the gas gradually gathers into lumps which will become the first stars. The increase of density leads to an increase in temperature, and when the temperature in the center of the star becomes sufficiently high (millions of degrees), the protons move so fast that it becomes possible to overcome their mutual electric repulsion and to fuse them into helium nuclei releasing a tremendous amount of energy in the form of electromagnetic radiation and neutrinos. This constitutes the simplest case of thermonuclear fusion.

Besides their role as sources of energy, the stars act as alchemical furnaces that produce elements heavier than hydrogen and helium (Fig. 4). The entities called elements in physics and chemistry constitute the bricks or components of a complicated "lego" set from which Nature makes more complex bodies. Elements are distinguished by the electric charge of their atomic nuclei.

Big Bang   Cosmic Rays   Light Stars   Heavy Stars   Supernovae

H																	He
Li	Be											B	C	N	O	F	Ne
Na	Mg											Al	Si	P	S	Cl	Ar
K	Ca	Sc	Ti	V	Cr	Mn	Fe	Co	Ni	Cu	Zn	Ga	Ge	As	Se	Br	Kr
Rb	Sr	Y	Zr	Nb	Mo	Tc	Ru	Rh	Pd	Ag	Cd	In	Sn	Sb	Te	I	Xe
Cs	Ba		Hf	Ta	W	Re	Os	Ir	Rt	Au	Hg	Tl	Pb	Bi	Po	At	Rn
Fr	Ra																

La	Ce	Pr	Nd	Pm	Sm	Eu	Gd	Tb	Dy	Ho	Er	Tm	Yb	Lu
Ac	Th	Pa	U	Np	Pu	Am	Cm	Bk	Cf	Es	Fm	Md	No	Lr

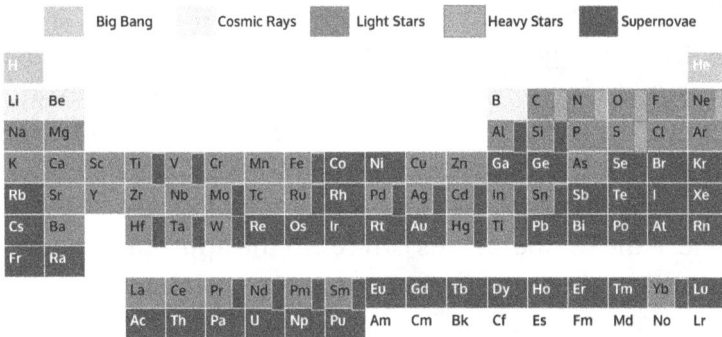

Fig. 4. A version of the periodic table indicating
the origins of the elements. All elements
above Pu (plutonium) have very small life times.

The types of atoms are listed in the Periodic Table and each element is marked by its own symbol: hydrogen as H, helium as He, carbon as C, oxygen by O, and so on. The atoms of a given element are composed of a tiny positively-charged nucleus and negatively-charged electrons bound to the nucleus by the electrostatic force. Electric charge is quantized which means that the charges of nuclei correspond to an integer number of electron charges. The electric charge of the nucleus of a given element is determined by its place in the Periodic Table. For instance, hydrogen (H) is in the first place and its nucleus has charge $+e$, carbon (C) occupies place number 6 and its nucleus has charge $+6e$ and so on.

Atoms can bind together to create complicated arrays. These may be molecules of various complexities or solid bodies. In doing so they exchange their electrons. Every complex structure created in this way has its own threshold of stability, characterized by the temperature at which it falls apart. The nuclei of the elements have the highest degree of stability, and stellar temperatures are required to change them. It was this requirement that doomed the alchemists' hopes of transforming "base" metals into gold using only chemical reactions: the amounts of energy involved in such reactions are insufficient to change the structure of the atomic nuclei which determine the character of the elements. However, such processes do become possible in the interiors of stars, where the temperatures can reach the appropriate magnitude to allow the fusion of lighter nuclei into heavier ones (nucleosynthesis).

Nucleosynthesis proceeds in stages because the heavier the nuclei the higher the temperature needed for fusion to take place. In order to fuse, the nuclei must overcome their mutual electrostatic repulsion, and the energy associated with this is proportional to the product of the charges of the colliding nuclei. So the heavier the nuclei the greater the energy needed for them to approach each other sufficiently closely for the strong nuclear attraction to overcome the electrostatic repulsion.

Let us consider in more detail the formation of carbon, the element that serves as the backbone for all biochemistry (Fig. 5).

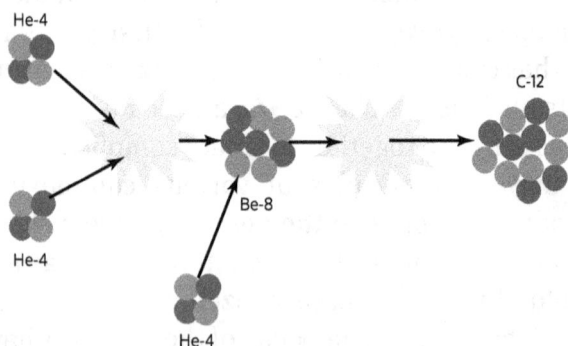

Fig. 5. The sequence of nuclear reactions leading from helium to carbon.

We will see that carbon owes its abundance to a very curious mechanism. As mentioned above, the first state of nucleosynthesis is a conversion of hydrogen nuclei into helium ones. The pressure of the radiation emitted in this process prevents the star from collapsing. However, once a significant part of the hydrogen is burned out, the pace of the fusion slows down and the radiation pressure that keeps the star at equilibrium decreases. As a result, the star core collapses. This leads to a rise in temperature and the collapse will continue until the temperature reaches the threshold beyond which fusion of helium nuclei is possible. But now there is a catch. Two helium nuclei fuse into beryllium-8, but the latter nucleus is unstable and decays back into helium. The curious fact (one of the many which we encounter when considering the relationship between cosmological processes and a possibility of our existence) is that the lifetime of the Be-8 nucleus is untypically large, about 100 times larger than typical decay times in other nuclear processes. In physics such a situation is called resonance. This long lifetime increases the probability of a triple collision when the third alpha particle (a helium nucleus) collides with the Be-8 nucleus and fuses with it, producing a stable nucleus of carbon. Since direct triple collisions are unlikely in comparison with those of pairs, a non-resonant fusion of three alpha particles into one carbon nucleus would not produce much carbon. The famous astronomer Fred Hoyle had predicted the existence of this resonance on purely "anthropic" grounds before it was actually observed. Hoyle thought that this was the only way to explain the abundance of carbon in the universe.

Some carbon nuclei can further fuse with additional helium to produce oxygen. This reaction goes along with a release of energy, but is not resonant. This creates a situation in which stellar nucleosynthesis produces significant amounts of carbon and oxygen, but not much of the heavier elements.

Here we first encounter the term "anthropic principle" (discussed later in detail). The connotations are clear: to produce carbon—which is so vital for life—Nature has to pass through an extremely narrow bottleneck. Indeed, the existence of complex structures such as organic molecules critically depends on the abundance of carbon; and the latter in turn hangs on the existence of resonance in the three-body collision of alpha particles, as well as on the absence of a similar resonance for the creation of oxygen to prevent a subsequent conversion of carbon into oxygen. Clearly, the existence of such finely-tuned resonances strongly favors life. And, in fact, this is just one favorable circumstance among many.

So we have learned how stars produce chemical elements, the building blocks for more complicated material systems. This production, however, would serve no purpose without a mechanism that enables the elements to emerge from the interior of the stars where they cannot form any stable structures since the temperatures are too high. The solution to this problem is the fact that old stars violently explode and become novae and supernovae. When most of the nuclear fuel is burnt out the temperature drops, thus breaking the balance between the forces keeping the star at equilibrium. The temperature decrease leads to a drop in the radiation pressure, which cannot longer resist the force of gravity that is pulling the outer layers of the star inward. The outer layers collapse and collide with the dense inner core, producing a shock wave that propagates outward through the stellar material igniting in it the thermonuclear fusion processes. As a result, the star explodes, dispersing the fusion products into outer space.

In fact, the supernovae bursts serve an additional purpose, since the formation of the heavier elements closer to the end of the Periodic Table occur not before, but during such bursts.

So stars are born and die; some of them explode as they die, thus passing their products onto the next generation of cosmic bodies. This takes time, a lot of time, billions of our earthly years. Our Sun with its planetary system abundant with heavy elements is believed to belong to the fourth generation of stars. The stuff of our bodies is made of stardust.

The next stage of complexity is reached when atoms of the elements create molecules. Molecules are much more fragile than atomic nuclei and individual atoms; the degree of fragility increases with their complexity. No chemical compounds can survive in stars, since they require environments where temperatures are much lower. Complex molecules may exist on comets or even in interstellar space, but at such low temperatures they cannot react with each other. In order to survive and to function, they require conditions which may exist only on planetary bodies.

DAY TWO:

The young Earth before the advent of life. The formation of the atmosphere and oceans. Water as a unique substance

Our solar system was formed about 5 billion years ago as determined from the study of meteorites, which are believed to be its oldest objects. The age of the Earth is estimated to be 4.6 billion years. The oldest rocks on the Earth, formed when the planet cooled down from a molten state, are dated at 3.8 billion years old.

There are two kinds of planets in the solar system: the so-called gaseous giants (Jupiter, Saturn, Uranus, Neptune) and planets with a solid surface (Mercury, Venus, Earth, and Mars and Pluto). The planets of the second group are composed of relatively heavy elements such as carbon, silicon, and iron. The elements whose presence is vital for life as we know it include carbon, nitrogen, oxygen, phosphorus, and hydrogen. Living organisms also utilize calcium (bones, etc.), iron (blood) and many other elements, though in small quantities.

Among these elements carbon occupies a particularly important position due to its rare ability to form long chains which serve as a backbone for molecules of practically unrestricted complexity. This property comes from

109.5°

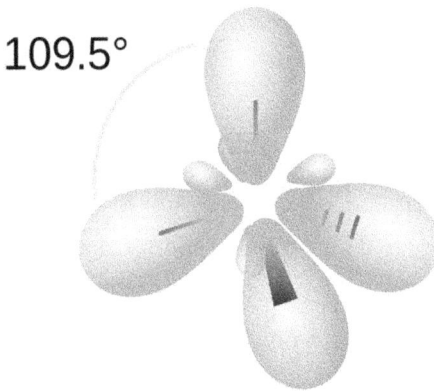

Fig. 6. A schematic representation of one of the valence
configurations of carbon atom. The big balloons represent
electronic clouds which the carbon atom can share
with other atoms. By overlapping with similar clouds
of other elements, they form strong chemical bonds.

the fact that carbon atom has four valence electrons: two
in $2s$ and two in $2p$ configurations whose energy levels
are very close to each other. This gives carbon atoms
an ability to form different chemical bonds of different
type (see Fig. 6, 7)—a property necessary for formation
of complex chemical structures (Fig 8). As discussed
in the previous chapter, the self-organization of matter
into complex molecular structures cannot be taken for
granted: it is conditional on the presence of large quantities of carbon.

The only other element capable of producing long
polymer chains is silicon (Si). (One can find descriptions of silicon-based life in science fiction; perhaps it
exists somewhere in the Universe.) However, there are
important differences between the chemistry of carbon
and silicon. For instance, the simple molecule carbon
dioxide CO_2 is a gas, but its silicon analog SiO_2 is a solid.

Methane (CH$_4$)	Ethane (C$_2$H$_6$)	Ethene (C$_2$H$_4$)
Tetrahedral (single bond)	Tetrahedral (single bond)	Planar (double bond)

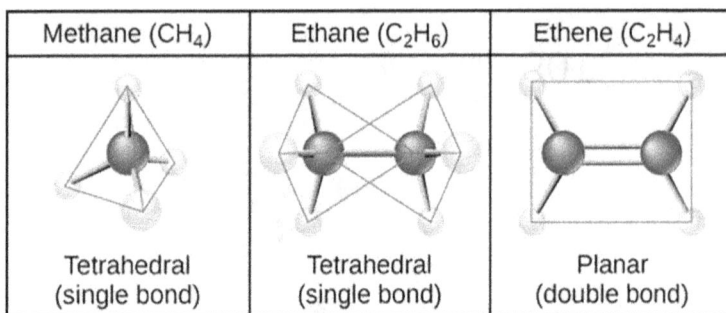

Fig. 7. Different types of carbon covalent bonds.

Hence carbon dioxide, being mobile, enables it to be used by plants, which would be difficult if it were a solid like SiO$_2$.

Whenever we look for the possibility of life we first look for water. Water serves as a solvent providing a medium for chemical reactions. Dissolving many types of molecules and transporting them to reaction sites, it nevertheless preserves their integrity. Hence life needs water in its liquid state, which can exist only in a certain range of temperatures between freezing and

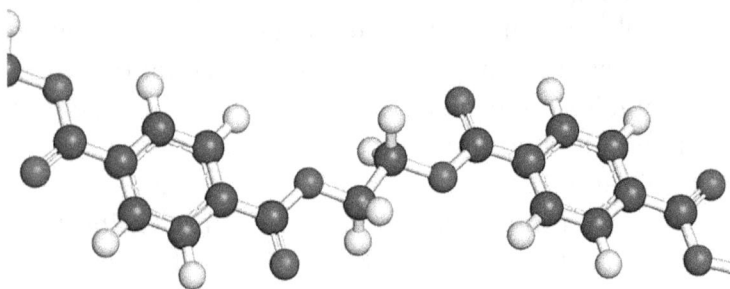

Fig. 8. A typical picture of a polymer with various atoms attached to the backbone formed by carbon atoms depicted as black balls.

boiling: the exact range depends upon the atmospheric pressure, but it is around 100 degrees Celsius. As astronomical observations suggest, temperatures on planet surfaces can vary from close to absolute zero (Neptune and Pluto) to one thousand degrees (Mercury). Temperatures on Earth fall within the special interval where water exists in all three of its states: solid (ice), liquid, and gas (water vapor) making it almost unique among all other natural substances on Earth. This property gives rise to the hydrological or water cycle providing water for terrestrial ecosystems which makes life on land possible.

There other quite special properties of water such as high specific heat, low viscosity, high surface tension and tensile strength which make it indispensable for life. "...water, by its *own powers*, delivers itself to the land, erodes the rocks, provides the minerals for terrestrial ecosystems, rests in the soil, and transits as needed through the soil to the roots...it is also by water's *own powers* that it raises itself and its precious cargo of minerals from the root to the leaf, enabling the production of life-giving oxygen trough photosynthesis".[8] These unique properties of water[9] originate from the so-called hydrogen bonds with which H_2O molecules may effectively interact with each other and with other molecules such as proteins and nucleic acids. These bonds determine the three-dimensional structures of these biologically important molecules and hence their biological functions. Hydrogen bonds are relatively fragile, which makes them highly flexible. The relative ease with which these bonds connecting the different strands of DNA spirals can be disconnected and reconnected again plays a leading role in the machinery of living cells (Fig. 9).

The high specific heat of water makes the oceans a source of the climate stability since large quantities of water is equally difficult to heat and to cool. The low viscosity, high surface tension and tensile strength provide water with ability to feed plants and trees by ascending along the narrow capillaries inside of their trunks and leaves and rising to great heights. The same properties enable water to fill small cracks in rocks which enhances their erosion thus providing nature with minerals necessary for life.

It would take too long to explain the remarkable properties of water, but let me describe just one. Everybody knows that ice floats on water, which means that it is less dense than water at the same temperature. This is

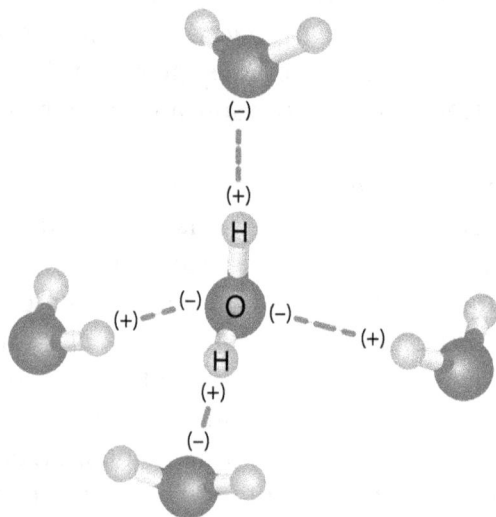

Fig. 9. The picture illustrates the mechanism of hydrogen bond formation between molecules of water. The bonds develop due to the Coulomb attraction between the predominantly negatively charged oxygen atoms (in black) and the predominantly positively charged hydrogen ones (in gray).

anomalous: most substances contract with cooling. In fact, below 4° Celsius fresh water expands when it cools, so that in winter the densest water is at the bottom of rivers and lakes. So the warmest water is protected from freezing by the outer layer of ice and the lower layers of cooler water. If it were otherwise, rivers and lakes at high latitudes would freeze to the bottom in winter, making the existence of complex life forms very difficult.

Here is a good overview on the emergence of the early atmosphere, stressing the importance of water:

> Following accretion into a large planet-sized object during the early years of the solar system, Earth's first major atmosphere was formed by the release of gases trapped in the restless interior, a process that still goes on today in volcanoes. These early years are marked by swirling oceans of hot magma that no longer exist today on any planet in our solar system. Extreme volcanism in Earth's early history occurred in response to this energetic motion of then-molten mantle material. As planetary material violently overturned, volatile gases from the interior, especially Carbon dioxide (CO_2), Carbon monoxide (CO), Hydrogen (H_2), Nitrogen (N_2) and water vapor (H_2O), were released, and accumulated in a gaseous surface layer that was trapped by gravitational forces. Radiation from the nearby Sun swept lighter gases as H and He away, leaving only heavier molecules in this early atmosphere. Chemical reactions in the hot surface layer formed other simple atmospheric compounds, such as methane (CH_4) and ammonia (NH_3). While far less abundant, the latter compounds are highlighted as they are key components of amino acids, which are the fundamental building blocks of life's proteins.

Note also that Oxygen (O_2), key to the survival of many forms of modern life, was not present in the early atmosphere.

Water vapor is the dominant form of gas released in outgassing, so much that Earth's early atmosphere became saturated with water, leading to an era of continuous rain on the planet. These rains contributed to cooling of the Earth's surface and the formation of the first oceans, which today cover two thirds of the surface. Another contributing process to the surface heat budget was the permanent removal of CO_2 from the atmosphere by the formation of carbonate rocks (consisting mostly of the mineral calcite, $CaCO_3$), which reduced the atmosphere's greenhouse potential and lowered the surface temperature [*This happened later on when the first algae appeared*—A. Tsv.].

The continuous rains and atmospheric CO_2 removal cooled our planet and created a surface where water could eventually be present in all three phases: vapor, liquid and solid. Ocean formation made Earth habitable for life at about 4000 million years ago, leading to today's highly complex organisms. The origin of life remains heavily debated, but the changes in organism over billions of years would not have been possible without surface conditions that persist today.[10]

DAYS THREE AND FOUR:

The emergence of life. Blue-green algae
change the atmosphere. The complexity
of even the simplest life forms

> *The origin of life appears ... to be almost*
> *a miracle, so many are the conditions which*
> *would have had to be satisfied to get it going.*
> —Francis Crick. 1981. Life Itself: Its Nature
> and Origin. New York: Simon & Schuster, 88.

The emergence of life marks a very dramatic increase in complexity. In his book *What is Life?* Erwin Schrödinger observed that assemblies of large number of atoms are stable in the face of statistical fluctuations. Hence such an increase was necessary to preserve the stability of biological structures.

The gap in complexity between even the simplest of living creatures such as single-cell organisms and the most complicated structures in inanimate matter is so vast that scientists still have no established theory about life's origins.

To get some idea about a minimally complex life form still capable of autonomous reproduction, scientists have built an artificial genome containing only the genes essential for reproduction, and implanted it into a receptive cell environment. The result was a "synthetic" bacterium dubbed JCVI-syn.3.0, whose genome is built from chemically synthesized oligonucleotides while the rest of the cell is taken from another bacteria.[11] The minimal genome is still very complex: it contains 531,000 base pairs forming a "library" of 473 genes. One may compare it to a book of 473 pages with each page containing about one thousand letters. This book contains instructions for the production of 438 proteins. As will be explained below, a genome is just a passive repository of information. Like a book sitting on the shelf, it is absolutely inert until someone takes it from the shelf and reads it. The reading and production functions are performed by the appropriate cell machinery whose presence the experiment takes for granted. The "Simplest Bacteria" is in fact much more complex than its genome, since it also contains the receptive environment taken from another cell. But even if we leave everything else aside and concentrate on the genome, even the simplest genome is not that simple.

This example of the bacteria with a synthetic genome illustrates two points. On the one hand, it emphasizes the enormity of the gap between life and not-life. The chance appearance of such a structure is highly improbable. On the other hand, it suggests that processes in living cells take place according to the same laws of physics we find in inanimate matter. Neither in the synthetic bacteria nor in more complicated microorganisms do scientists find any sign of a special "life force." This adds to the mystery of the whole—which is an extremely sophisticated construction where different parts work together towards a common goal. This brings to light another aspect of the problem. The genome will not work without the receptive environment of the cell and this environment will fall apart without the genome. This conundrum is the chicken and egg problem which has always bedeviled research on the origin of life.[12]

Living cells became areas where the second law of thermodynamics appears to run backwards: since the cells increase order, entropy actually decreases instead of increasing. But this does not present decisive evidence that the law is broken. In open systems, entropy can decrease, since this decrease is compensated for by an increase in the entropy of the environment. Nevertheless, such processes cannot be considered as typical, which is evident from the fact that life is a rarity and within the solar system is likely to exist only on our Earth.

The first life forms appeared very quickly once the Earth cooled down. The fossils of microbes using photosynthesis are found in Australian rocks formed 3.5 billion years ago: that is just 300 million years or so after the formation of the Earth's solid crust![13]

This might be compared with other time scales for the development of eukaryotic single-cell and multicellular organisms. Fully developed eukaryotes are believed to appear from two billion to one billion years ago. Multicellular organisms emerged about 600 million years ago during the Cambrian period.

The disparity between these time scales is striking in view of the fact that the structural gap between the raw chemicals naturally available on the pre-life Earth and living cells is much greater than the gap between single-cell organisms and complex life forms. In view of this, some scientists advocate the so-called "panspermia" hypothesis, according to which life arrived on Earth from elsewhere. It is not clear how much time one can gain this way. The problem is that life cannot be formed without heavy elements and, as we have seen, this process took billions of years, so that the material constituents of life could not have arrived on the scene much before the formation of the solar system. So the panspermia hypothesis may conceivably gain a couple of billions of years, but would it be enough to cover the gap? We do not know, since we are not yet in a position to obtain quantitative estimates of the timescales involved in the evolution of life forms to test the consistency of the theories.

Earth's atmosphere at the time of the first microbes contained no oxygen in a free form. It was opaque and the sky was covered with dark clouds. It took the microbes more than a billion years to fill the atmosphere with oxygen and make it transparent to visible light. These microbes, which are believed to be similar to modern blue-green algae, fed off the carbon dioxide, emitting pure oxygen much like modern plants do. The process of atmosphere formation was very gradual since most of the oxygen generated by the algae was consumed by the oxidation of rock formations. Only when the rocks were saturated with oxygen did the atmospheric content started to change, and around one billion years ago the free oxygen started to stay in the air.

So life has transformed the planet and by doing so created the conditions for its own progress. This is an early example of how life itself can change the process of evolution by changing the environment. This change was accomplished by very primitive single-celled organisms whose cells lacked a membrane-bound nucleus (prokaryotes). Without this transformation life could not have developed into more complex forms.

The first single-cell organisms whose cells possessed a nucleus (eukaryotes) appeared about two billion years ago. Their appearance signified the next enormous step of increasing complexity, since eukaryotic cells are typically much larger (by a factor of 1000 in volume) than those of bacteria. Their cells are highly structured and compartmentalized, being held together by a cytoskeleton structure. This compartmentalization is absent in even the most advanced prokaryotes. The compartments (organelles) are separated by highly structured porous membranes. The content of prokaryotic cells is a viscous solution so that macromolecules (proteins and nucleonic acids) can freely diffuse. This freedom is denied them in the eukaryotes, and they reach their "working places" using a complicated transport system. This high level of organization is reflected in low entropy, which can be demonstrated experimentally by piercing the outer membrane of prokaryotic and eukaryotic cells. In the former case the contents of the cell just spill out, but in the latter one they do not. This difference in entropy density constitutes the most fundamental difference between prokaryotes and eukaryotes.

Eukaryotic DNA is divided into separate bundles (chromosomes), with each bundle containing many genes. Eukaryotes are the first organisms to pioneer sexual reproduction.

There are a few key facts about life that I would like to emphasize here. The first is that life is fragile. The second is that it develops very slowly; the progress of life from the simplest single-celled organisms to humans has taken billions of years, a timescale comparable to the age of the Universe. Hence, as we have seen, the long duration of the evolutionary process requires very stable conditions on the cosmological timescale. The possibility of such conditions prevailing depends on the character of physical laws, which must allow for highly stable planetary orbits and protection from collisions with other cosmic objects. The duration of the evolutionary process also explains the apparent paradox of the large size of the Universe.

It turns out that even if the phenomenon of life is unique and does not occur anywhere else in the Universe, an enormous Universe with myriads of stars and galaxies is still needed to sustain it. The Big Bang theory allows us to relate the mass of the observable Universe M_u and its age t_u:

$$M_u \sim 10^5 (t_u/1sec) M_{sun}$$

(This formula states that the mass of the observable Universe measured in the Sun masses is approximately 100,000 times greater than its age measured in seconds.) This formula follows solely from the theory of gravity and would be valid even if there were no other forces in the Universe (Barrow and Tipler 1986, 384–85). Although we do not know how to calculate the time necessary for life to evolve, there are reasons to believe it is billions of years. Add to this the time necessary for stellar nucleosynthesis and we get something close to the current estimate of 13.4 billion years. Since one year contains approximately 31 million seconds, we conclude that the Universe supporting life on at least a single planet must contain more than 10^{17} (100,000 trillion) stars of the Sun's class—which makes it quite big.

We have good reason to believe that during the last 600 million years the changes in the average temperature on the Earth's surface have not exceeded several degrees. So conditions have remained stable, with the exception of relatively brief periods of catastrophic change that caused great extinctions of life forms, the best known of which is the one that led to the extinction of dinosaurs 65 million years ago. This disaster was most probably caused by a collision with an asteroid or a comet which fell somewhere in the region of the modern Yucatan peninsula. The massive explosion that followed filled the skies with debris, thus depriving the Earth of a substantial part of solar energy.

Asteroids are abundant in the solar system and one may wonder why such collisions do not occur more frequently than one in a hundred million years. There is a reason for that: the asteroids and comets are deflected from Earth by the giant planet Jupiter, which serves as a kind of a cosmic vacuum cleaner. Without this giant planet with its strong gravitational force the Earth would be in tremendous danger. To appreciate the scale of possible destruction one should look at photographs of Comet Shoemaker-Levy 9 striking Jupiter in 1994.[14]

Let me return now to the complexity of life itself, the issue raised at the beginning of this chapter. To say that even simple living organisms are very complex does not really do justice to the subject. To appreciate what we are talking about we need details. I emphasize that the picture presented below is valid for all known life forms, including the most primitive.

The impression one gets from watching animations of working living cells (such as can be found at www.wehi.edu.au) brings to mind the workings of an automated factory. The main functions of the cell, whether related to pure reproduction, metabolism or (in multicellular organisms) to something else, are performed through an elaborate cooperation between several departments. This cooperation is supported by constant communication—there are messenger molecules scurrying back and forth. There is a department that contains instructions for the work the cell can perform. This information is written in a code on enormously long molecules called DNA (they can reach hundreds of meters in length). Rungs of the DNA molecule (they are called base pairs; see Fig. 10) serve as the letters of a four-letter alphabet. Pieces of DNA containing particular instructions are called genes; each gene contains hundreds or even thousands of letters. Genes can be compared with blueprints. The human genome contains about 3 billion letters (base pairs) (Fig. 10).

The DNA library with its contents is absolutely inert just as any library is: to make the stored information useful, somebody must read the particular piece of text needed at a given time. The process of reading goes through several stages. First, there are special proteins which select those segments of DNA which should be read at a given time. Then the extracted information is transmitted from one part of the cell to another and also used by the cell machinery for protein synthesis, as is explained below. Special messenger molecules (mRNA)

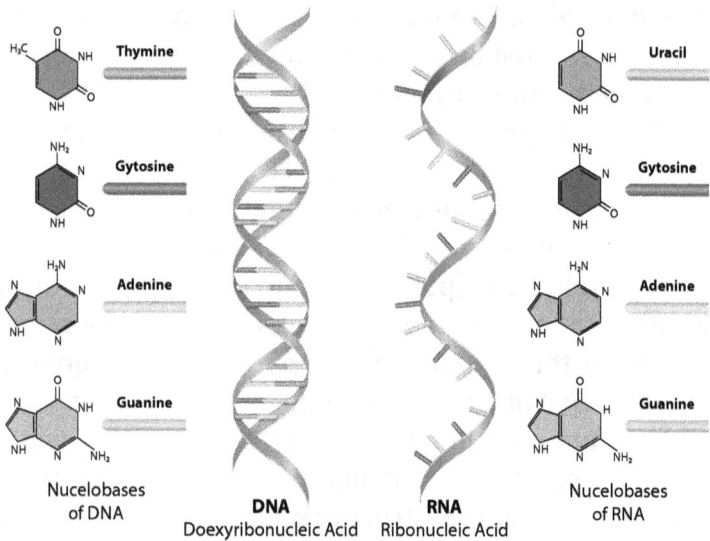

Fig. 10. Pieces of DNA and RNA molecules. The double helix of DNA serves as a support for the text encoded in the connecting molecules of four different types (thymine, cytosine, adenine, guanine) performing the role of letters of the DNA alphabet. RNA molecule contains uracil instead of thymine.

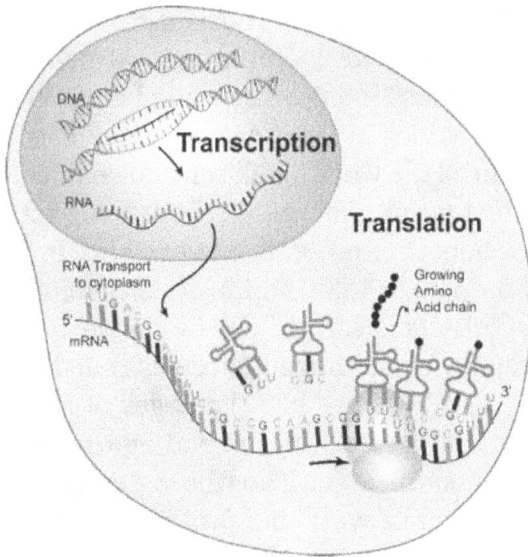

Fig. 11. The transport of information inside a living cell.

are directed by other molecules (enzymes) to the relevant genes to copy the pieces of the blueprint and carry them to the production department (Fig. 11).

Information stored in DNA is not just inert. It is also written in a code and hence cannot be used directly for production. This fact is absolutely remarkable and constitutes yet another great mystery about life because the introduction of code presents yet another chicken and egg problem. Coded information is useless without reading machinery and the reading machinery cannot function without coded instructions. So the two parts must be present simultaneously, but both are quite complicated. And, as far as we know, there are no physical laws that would give preference to one code over another. The situation with life is unique: we are not aware of any other cases of using codes in nature apart from their use in biological organisms.

The production process consists of making proteins from biologically active amino acids (there are only twenty such acids among a large number of structurally similar ones which are not active), since it is the proteins that do all the work in the cell and enable it to function. The text written on mRNA needs to be activated in order to connect amino acids into proteins. Proteins are formed from 100 to 1,000 amino acids chained together. Protein synthesis is carried out by the so-called transport RNA (tRNA) and huge molecules called ribosomes (see Fig. 12). tRNAs are like the pages of a dictionary; they carry amino acids on one end and several letters of DNA code on the other. The ribosomes are absolutely necessary. To make workable proteins the amino acids must be connected end to end. This process is directed by ribosomes which prevent one acid molecule from

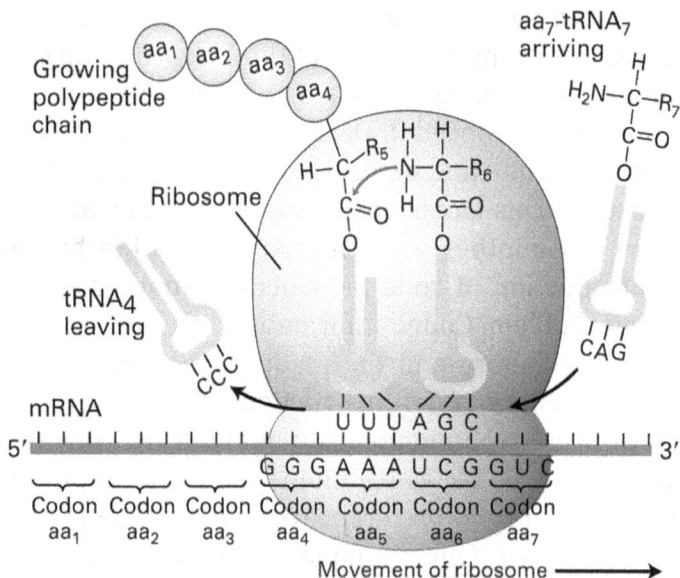

Fig. 12. Protein synthesis.

connecting to the side of another. A ready protein molecule undergoes a complex folding to take on the specific shape that determines its activity. Proteins with side chains of amino acids are not functional (Fig. 13).

Naturally, the process of protein synthesis requires a controlled supply of energy. For this purpose cells use ATP molecules (a kind of nucleotide). These molecules provide tiny amounts of energy that are exactly right for biochemicals: the amount of energy produced is just enough to perform the task, but sufficiently small not to damage the cell. The Krebs cycle, in which ATP molecules are produced, requires eight enzymes to operate; the absence of any one of them will make it nonoperational. Enzymes are huge molecules made of hundreds of amino acids. This complexity arises from the highly specific functions they perform.

The entire process of cell operation resembles the execution of a computer program. This analogy goes only so far, however; it would be wrong to think of DNA as static software and the rest as hardware. Cells react to the environment and reprogram themselves accordingly.[15]

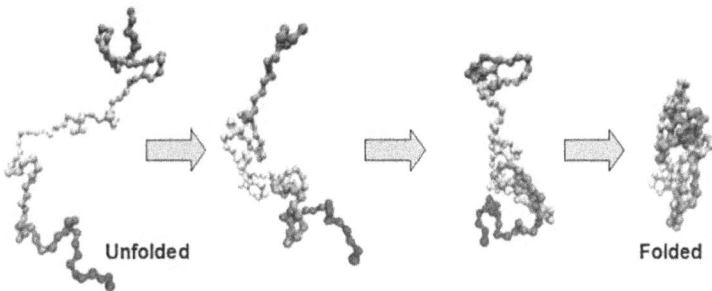

Unfolded   Folded

Fig. 13. Protein folding.

I hope I have managed to give the reader a glimpse into this long and elaborate process whereby the word encoded in DNA literally becomes flesh (proteins). Now we can only exclaim with Dr. Francis S. Collins, head of the Human Genome project: "How marvelous and intricate life turns out to be! How deeply satisfying is the digital elegance of DNA! How aesthetically appealing and artistically sublime are the components of living things, from the ribosome that translates RNA into protein, to the metamorphosis of the caterpillar into the butterfly, to the fabulous plumage of the peacock attracting his mate!" (Collins 2006, 107).

DAY FIVE:

Multicellular organisms and the creation of the internal world

> *"The complexity of living bodies has to be present either in the material [from which they are derived] or in the laws [of their formation]."*
>
> —Kurt Gödel[16]

Evolution

In any discussion of life it is impossible not to comment upon the idea of evolution. The term has become heavily loaded due to the recent ideological battles waged by religious fundamentalists on one side and the "new" atheists on the other.

In any discussion of evolution there are a few things about which one has to be very clear or the discussion will lead nowhere. First, one has to distinguish between the facts and the theories aimed at explaining the facts. To criticize the theories is not the same thing as to deny the facts. Second, one should be aware that there is not a single theory of evolution but many, and the arguments among them are not yet over.[17] The life sciences are not yet in a position to present a comprehensive picture of life and we are still learning *fundamental* facts that dramatically alter our understanding of the process. For this reason it is difficult to make generalizations, and one should be very cautious in trying to extract philosophical conclusions from the life sciences as presently constituted.

Let us list some important facts that are widely recognized in the science community. The most important fact is that life has had a long history, during which various life forms emerged and disappeared. This is demonstrated by the fossil record, which stretches back to the rapid appearance of primitive life forms 3.5 billion years ago. We also have ample evidence to believe that all life forms are related to each other, since they share many features on the microscopic level. The most important of them is the genetic code, briefly described in the previous chapter. Hence we can trace relations between different life forms by looking for similarities in their genetic material. In particular, all multicellular organisms share a common set of the so-called "master" genes responsible for the building of their bodies. The presence of these common features makes implausible the belief that each kind of living creature was created independently from scratch. Another widely recognized fact is that, despite their incredible sophistication, living forms cannot be regarded as "perfect." All these facts provide evidence against a naïve creationism, but neither do they by themselves guarantee the triumph of a naïve naturalism leaving room for other options.

Taken together these facts constitute evidence for the *evolution of life forms*. However, they do not in themselves contain an explanation of the evolution. This is the subject of frequent popular confusion. Even if the above facts are accepted as established ones, one is still left with a myriad of questions that must be answered by *a theory of evolution*, whatever it may be. What mechanism brings about the emergence of new species? How does one form of life give rise to another?

Does evolution require new laws beyond the ones we know in physics? Could it be that there are special forces present only in living beings which affect the evolutionary process? Is it a blind process or is it driven towards some purpose—perhaps, even under divine guidance? To what extent can living creatures themselves influence the process?

Answers to each of these questions will heavily affect any philosophical conclusions drawn from the facts of evolution. The answers best known from popular literature are given by the Darwinian (or rather neo-Darwinian) Synthetic Theory of Evolution (STE). The formation of this theory was completed by the late 1950s. The theory tacitly assumed (as Darwin himself believed) that life started from "simple beginnings." STE also stated that evolution is gradual and proceeds by an accumulation of small changes due to natural selection.[18] The main mechanism of change was chance mutations of the genome, selected for their adaptational advantage. The object of selection was the phenotype, that is, the set of an individual's observable characteristics resulting from the interaction of its genotype with the environment.

STE was formulated prior to the discovery of DNA, the mechanism responsible for the storage of hereditary information *common to all life forms.* No wonder then that its main tenets have been proven wrong.[19] However, the legacy of STE has survived in the form of a myth whereby "natural selection" emerges as the universal key capable of opening any door. Everything became "evolutionary" from phycology to physics. As I will discuss in the section on the Anthropic Principle, this myth has found its way, in the form of Multiverse Theory, even into fundamental physics. The general idea is that complex forms may emerge by themselves without any creative thought or plan behind the process. This belief constitutes a cornerstone of the ideology of New Atheism. However, such a conclusion remains absolutely unwarranted. Any natural change is contingent on the structure of the laws of physics, whether it is the evolution of life forms or the self-organization of inanimate matter. Although vastly more complicated than inanimate matter, life—at least in its microscopic forms—can be considered as just another step in the process of the self-organization of matter, which has been going on in the Universe right from the beginning.

In the world of inanimate matter, more complicated systems emerge from less complicated ones not accidentally, but according to a logic encoded in the laws of physics. Hence this process cannot be considered as unguided. New and more complex forms emerge without any Darwinian selection for the simple reason that things that are not alive cannot be deselected by death. There is no "survival of the fittest" for things that are not alive. For instance, quarks gave rise to protons and neutrons; together with electrons they made atoms; atoms gathered into molecules and so on and *this was not an evolution in the Darwinian sense*. One cannot ascribe this process to pure chance, since it occurs according to the laws of physics, which in the ontological sense precede all material processes, as I have stressed in the Introduction. Most of the processes in living organisms also follow the laws of physics and chemistry and hence their outcome is *unambiguous*.

The Law—Logos—precedes matter, and in that sense *order is always preceded by more order*, just as an idea in the mind of a sculptor precedes the material incarnation of the sculpture. Any confusion about the process arises only because the preceding order is not directly visible, but exists in the realm of ideas. There is no such thing as "Design out of Chaos without the aid of Mind".[20] To recognize this, one has to see this realm as no less real than the realm of material substances. The other reason for confusion is the dogma that natural laws cannot in themselves contain any plan or goal. However, the existence of such plans follows, among other things, from the fact that self-organization is conditional of the special structure of laws of physics. We can see this clearly in the case of inanimate matter, but the same holds true in biology.

Although self-organization exists both in inanimate and animate nature, it is achieved by completely different means. In the inanimate world stable structures emerge as local free energy minima. For living cells, however, this is absolutely not the case. No arrangement of base pairs in DNA is energetically preferable to any other. As a consequence, the difference between gibberish and meaningful text is not decided by the gene itself. Left to its own devices the gene library would not contain any useful information at all. Such a state would correspond to the maximal entropy of the gene. However, genes do not exist in isolation. Their content is selected by the ability to initiate the synthesis of an appropriate protein in the production department. The correct text gives rise to an action that supports the work of the entire system. In terms of physics, a functioning cell can be described as being in a non-equilibrium steady state. Such a state can exist only in an open system in the presence of a constant supply of low entropy from the outside and the removal of high entropy products (metabolism).

The discoveries in genetics, microbiology, biophysics, and embryology over the past thirty years have provided us with an enormous amount of concrete detail that required a new synthesis to replace STE. It appears that such a synthesis has not yet been achieved at least to everybody's satisfaction. The contemporary state of evolutionary biology is characterized by a pluralism of models where natural selection is regarded as just one among many mechanisms that drive evolution. It has been realized at last that positive selection can work only in very large populations (of prokaryotes, for example), since in small populations harmful mutations which probability always far exceeds the probability of beneficial ones, will kill the population before any beneficial mutation occurs. Hence many significant evolutionary events could not be positively selected (that is selected for their adaptational advantage). This is the central theme of a remarkable book "The Logic of Chance" by the evolutionary microbiologist Eugene Koonin (2012) who emphasizes the leading role of chance in the history of life. In his attempts to find natural explanations for events that would otherwise appear as miraculous (such as the appearance of eukaryotes), he even invokes the newly fashionable Multiverse theory.

Multicellular Organisms

Let me return now to the main subject of this chapter, namely, the evolution of multicellular organisms. It seems that in this area a synthesis has been achieved in the form of Evolutionary Development (Evo Devo) theory. Evo Devo has absorbed the discoveries of both genetics and embryology, which together reveal a logic and order underlying the generation of animal form. And, I dare say, the spirit of this logic is dramatically different from the spirit of STE. Evo Devo strongly emphasizes the universal features common to all multicellular organisms. "For more than a century, biologists had assumed that different types of animals were genetically constructed in completely different ways. ... But contrary to expectations of *any* biologist, most of the genes first identified as governing major aspects of fruit fly body organization were found to have exact counterparts that did the same thing in most animals, including ourselves. This discovery was followed by the revelation that the development of various body parts such as eyes, limbs, and hearts, vastly different in structure among animals and long thought to have evolved in entirely different ways, was also governed by the same genes in different animals" (Carroll 2005, 9).

The observed diversity of forms in the animal kingdom "arise from evolutionary changes in where and when genes are used, especially those genes that affect the number, shape, or size of a structure ... this has created tremendous variety in body designs and the patterning of individual structures" (Carroll 2005, 11).

In *The Origin of Species* Charles Darwin urged his readers to see the grandeur in his vision—in how, "from so simple a beginning, endless forms most beautiful and most wonderful have been, and are being, evolved" (Darwin 1859, 490). The grandeur is certainly there, what is lacking is the "simple beginning". As we have seen in the previous chapter, even the simplest imaginable bacteria are complex organisms. And as far as multicellular organisms are concerned, they started with an immensely complex hierarchical apparatus of various genes which largely determined the subsequent evolutionary process.

The main conclusion we can derive from Evo Devo is that the formation of new species is directed and determined primarily by the internal properties of the organisms, for the most part by the laws governing the functioning of the genetic apparatus. New species of organisms emerge not by means of gradual small changes, but by abrupt and significant mutations. The genes of a given organism are arranged in so-called polymorphic and monomorphic groups. Genes in the first group mutate easily, and are responsible for differences between members of the same species, but do not change species. Darwin based his theory primarily on the readily available facts of such mutations (recall the finches) and hypothesized that the gradual accumulation of such changes might lead to grander ones. In fact, this idea has been proven wrong, because the changes necessary for the production of new species come from abrupt and revolutionary mutations of the genes of the second group. Monomorphic genes are very stable; if mutations occur, they are usually lethal. But the formation of new species comes from rare and dramatic mutations of this group of genes. Such mutations are due to the activity of genetic elements capable of moving within the genome, producing changes by attaching themselves to the DNA. The places where such attachment can occur are not randomly distributed in the genome. This fact heavily restricts the possible outcomes.

The discovery of the hierarchical structure of the genome has enabled us to explain why organisms with similar genetic content are so different. This difference comes from the different ways the master genes (see below) rule the whole. Every property or feature of an organism is determined not by separate genes, but by the entire genome whose functioning is regulated by an integrated system of switches located in the so-called "dark" portions of DNA.[21] The decisions to use or not to use a particular piece of information encoded in DNA is taken collectively by the entire genome; it is not an automatic process occurring independently of circumstances.

The highest place in the hierarchical structure of genome is occupied by the so-called master control genes. "Despite their great differences in appearance and physiology, all complex animals—flies and flycatchers, dinosaurs and trilobites, butterflies and zebras and humans—share a common 'tool kit' of 'master' genes that govern the formation and patterning of their bodies and body parts" (Carroll 2005, 9). There are segregation genes responsible for the division of the body into parts. There are homeosis (Hox) genes responsible for the qualitative specifics of body parts. There are master genes which produce DNA-binding proteins that switch on parts of genome responsible for the development of particular organs—Pax-6 genes for the eyes, Dll genes for parts that protrude from the body, Tinman genes for the heart, etc.—in all multicellular organisms. There are also tool kit members that signal pathways responsible for communication between cells. All these higher genes give commands to slave genes that perform the specific tasks.

While looking at the universal features of animal development, common to all creatures whose evolutionary pathways had diverged hundreds of millions of years ago, one cannot fail to recognize that this development follows the same *logical* pattern. "Just as in the construction of a building, where there is an order to the sequence of steps—the foundation is poured, the supporting walls and beams erected, the floors laid, major ducts placed, plumbing, electricity, drywall installed, etc.—there is an order in building animals, from the making of the basic body plan to the fine detailing of individual body parts. And, from *logic* [the italics are mine—A.Tsv.] of this order, we then understand how monstrosities result when the operation of a tool kit gene is damaged by mutation. When a step is omitted, all dependent steps are abnormal" (Carroll 2005, 106). The labor of construction is performed with tool kit genes, which are, like the tools of human constructors, used and reused for different tasks whose sequence is controlled by a system of genetic switches.

The large-scale divisions between multicellular organisms, classified in biology as *phyla*, were in all likelihood related to mutations in the master genes at the top of the hierarchy. These most important divisions, charting differences in body plans, preceded the development of more detailed differences internal to the individual phyla.[22] The discoveries that followed defy imagination. In particular, it has been found that species from different phyla (like flies and mammals, for instance) share the same genes for switching on the development of similar organs. According to STE organs such as eyes developed independently in different phyla, based on the fact that the eyes of the dragonfly, octopus, and mouse have completely different constructions. It turned out, however,

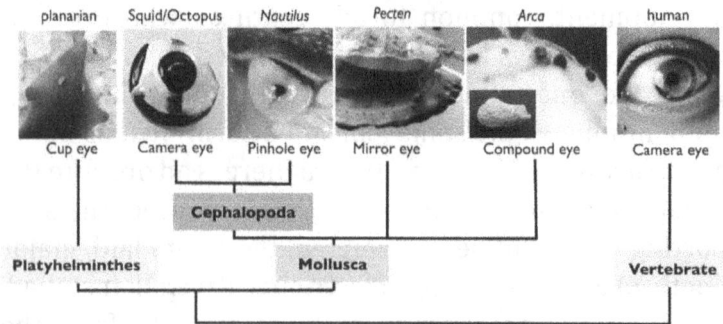

Fig. 14. Different types of eye.

that the development of the eye in organisms from different phyla is switched on by the same gene (*Pax-6*)! "Natural selection has not forged many eyes completely from scratch; there is a common genetic ingredient to making each eye type, as well as to the many types of appendages, hearts, etc. These common genetic ingredients must date back deep in time, before there were vertebrates or arthropods" (Carroll 2005, 72).

The Second Act of Creation

The first multicellular organisms (sponges, primitive mollusks, worms) appeared during the so-called Edicaran period about 570 mln. years ago. This relatively brief (about 13 mln. years) period signified an enormous leap in functional complexity of the organisms. It was, however, completely overshadowed by the subsequent event dubbed the Cambrian explosion (about 540 mln. years ago) during which most of the major animal body plans appeared including *Chordata* (vertebrates). According to MIT geochronologist Samuel Bowring the peak of the morphological innovation had spanned as little as 6 mln. years.[23] The net result was an enormous diversification of the animal kingdom. During the explosion at least sixteen novel phyla and about thirty classes first appeared in the fossil record. Animals with heads, guts, anuses, complex eyes of different kind, jointed limbs tentacles and many now familiar and then complete novel organs appeared. As a result the disparity of life had reached today's levels. Cambrian seas were full of animals of various forms, shapes, and sizes, occupying all possible ecologic niches. It is believed that the Cambrian explosion firmly established the fundamental structure of marine life which continues to this day.

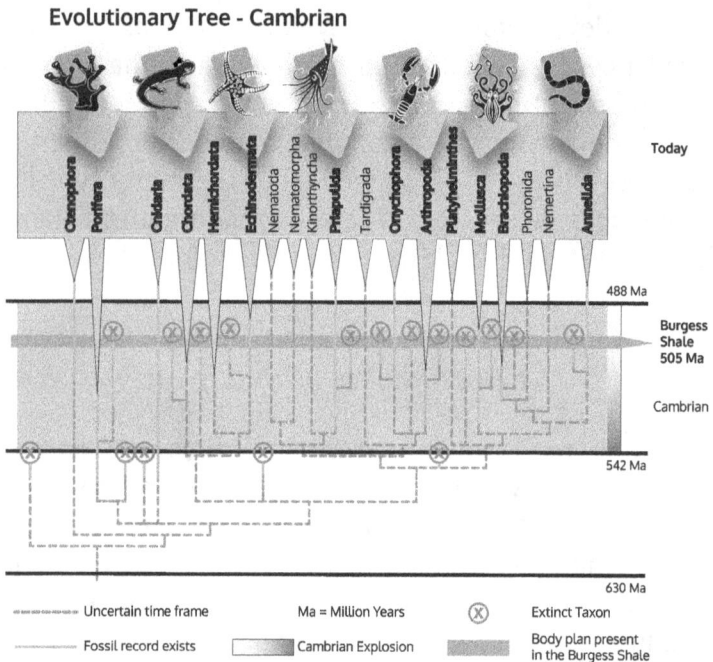

Fig. 15. Fossils of animal phyla found in the Burgess Shale.

Although the emergence of even the simplest life forms signified a colossal increase in complexity, we have no evidence for believing that the processes in living cells are governed by anything other than the laws of physics that govern inanimate matter. However amazing it may seem, the immensely complicated machinery of the cell is driven by the Coulomb force acting between electric charges and quantum mechanics which can be found in a graduate textbook. Non-equilibrium steady states are known to exist in inanimate systems, although the ones we know about are vastly more primitive than living cells. The fact that the complex behavior of microorganisms can be traced to such simple origins is, of course, mind-boggling; but one can still say that from the Big Bang to microorganisms *nothing had been created*. It is all the same physics: even with reference to primitive life one can perhaps say that it was, metaphorically speaking, *brought forth by the earth.*

However, the emergence of complex life forms brings with it something *completely new.* This novelty is the internal world of living creatures, the world of emotions, feelings, intentions, and, perhaps later, thoughts. With its appearance the Universe acquires a new dimension. Although creatures who possess all these always have a nervous system and a brain, one should not confuse these two areas of reality. One could well imagine and possibly even construct a machine that reacted to the environment in a way similar to, say, a fish. We have no reason to believe, however, that such a machine, which could be programmed to avoid destruction, will *suffer* from the damage inflicted on it. And it is difficult to believe that even such a distant relative of ours as a fish does not suffer. Why and how would a blind process of evolution produce an internal world of emotions, feelings, thoughts, and eventually consciousness, if one can perfectly well survive without them? What I call the "internal world" (IW) looks to the naturalists like a useless appendage, and they have no clue how to explain it.

It is my deep conviction that IW cannot be scientifically derived from previous events and should be taken as a fundamental or irreducible entity, one that cannot be derived from anything else. The emergence of IW can be compared to the appearance of an extra spatial dimension, and although one can imagine reducing the dimensionality of an object (say from three to two by flattening a metallic cube with a hammer), it cannot be increased. Therefore, the fact that IW is known to exist only in creatures with a nervous system and a brain does not make it a product of these two.

From the Cambrian Explosion to the Present

As I have mentioned above, the principal organizational structures of the animal kingdom were formed about 500 million years ago. No new animal phyla have emerged since then. One may notice certain analogies between the Cambrian explosion and the Big Bang: subsequent evolutionary history has been a realization of the potential encoded at the beginning of the process. This realization has been directed by the laws of evolutionary development which have been revealed to us by the successes of Evo Devo. As is the case with the laws of physics, these laws leave plenty of room for chance. Chance and necessity are intermingled so that the results of the evolutionary process cannot be predicted in detail (although looked at retrospectively, one can clearly see certain tendencies). Not every evolutionary branch develops more sophisticated creatures, but on the whole there is a tendency to produce larger and larger brains. It is curious to see how the leaders of the brain formation process emerge in different branches of the animal kingdom, as if to be tried and eventually abandoned until the proper ones are found. There was a time when the most intelligent creature was the octopus, which still maintains its leadership among invertebrates. However, its evolution stopped hundreds of millions of years ago. Dolphins are very intelligent animals, but they stopped evolving 30 million years ago. The animals who have continued to change and progress towards more complex brains are the apes. It is from them that we inherited our bodily form and organization.

DAY SIX:

Creation of mankind

Consciousness cannot be explained in physical terms since it is absolutely fundamental. It cannot be derived from anything else.
—Erwin Schrödinger[24]

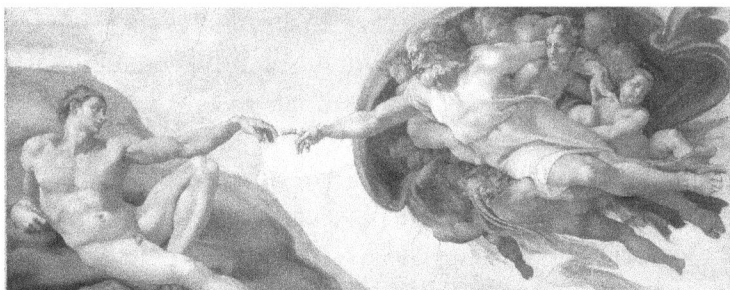

These words uttered by one of our greatest scientific minds bring us exactly to the point. Whatever the physical similarity between us and our ape ancestors, humans are fundamentally new creatures by virtue of having a reflective consciousness. This new element signifies an unsurpassable gap between us and the rest of the animal kingdom.

Our intelligence has truly revealed itself as a planetary force compressing both space and time. It has transformed our planet faster than any other force described in the previous pages. The "day" of creation associated with mankind is the shortest one in the succession of eons. During this day time became *history*.

The reflective consciousness has created culture and science as a part of it. The founding fathers of modern European science had been careful to exclude mind from the domain of science. A quote from Thomas Nagel:" Galileo and Descart made a crucial conceptual division by proposing that physical science should provide a mathematically precise quantitative description of an external reality extended in space and time, a description limited to spatiotemporal primary qualities such as shape, size and motion, and to laws governing relations between them. Subjective appearances, on the other hand—how physical world appears to human perception—were assigned to the mind, and the secondary qualities like color, sound, and smell were to be analyzed relationally... It was essential to leave out or subtract subjective appearances and the human mind—as well as human intentions and purposes—from the physical world in order to permit this powerful but austere spatiotemporal conception of objective physical reality to develop."[25]

Materialism with its claims to provide an encompassing explanation of the world has ignored this distinction and has been trying to incorporate consciousness in its scheme. However, despite all the hype and perpetual declarations of victory this attempts remain futile. The existence of mind and consciousness continue to present enormous problems to materialists.

These problems are multiple, and the first one is that they find no use for conscious mind.[26] According to materialist belief, mind cannot affect physical processes in the organism. If so, reflective mind brings no evolutionary benefits and should be deselected, which puts the materialists into embarrassing contradiction with one of their most sacred tenets—the theory of evolution. To disentangle this knot some of them even deny the very existence of conscious mind. Darwin himself did not belong to that category, as is clear from his argument with his friend and colleague Thomas Huxley, who thought that humans were some kind of automata. Darwin's views on the subject can be found in his book *The Expression of the Emotions in Man and Animals* where he had shown in great detail how the emotions of men and animals can and do express themselves in muscular movements.

Another difficulty for materialist science is related to the problem of knowledge. If human beings developed their mental capacities in the earthly environment how have we come to know so much about the Universe? Our ability to successfully perform various tasks well beyond the earthly environment such as space travel[27] is evidence of the reliability of our knowledge. But if such cosmic knowledge is indeed reliable, this means that our minds have a cosmic dimension and our reason has an affinity with the Reason who organized the Universe. In fact, as was stated in the Introduction, the natural sciences are based on the implicit assumption of the presence of such a Reason in the form of Natural Laws. Our ability to grasp these Laws signifies, as Socrates argues in *Phaedo*, that the soul has a close affinity with ideas (Forms).

The American philosopher Alvin Plantinga has drawn out some of the consequences of such a material-ist ("naturalist") point of view:

> From a theistic point of view, we'd expect that our cognitive faculties would be (for the most part, and given certain qualifications and caveats) reliable. God has created us in his image, and an import-ant part of our image bearing is our resembling him in being able to form true beliefs and achieve knowledge. But from a naturalist point of view the

thought that our cognitive faculties are reliable (produce a preponderance of true beliefs) would be at best a naïve hope. The naturalist can be reasonably sure that the neurophysiology underlying belief formation is adaptive, but nothing follows about the *truth of the beliefs* depending on that neurophysiology. In fact he'd have to hold that it is unlikely, given unguided evolution, that our cognitive faculties are reliable. It's as likely, given unguided evolution, that we live in a sort of dream world as that we actually know something about ourselves and our world. (Plantinga 2007)

Or, in the words of other contemporary philosopher, Thomas Nagel: "Evolutionary naturalism implies that we should not take any of our convictions seriously, including the scientific world picture on which evolutionary naturalism itself depends". Ironically, in its attempts to naturalistically explain consciousness the materialistic science has finished with undermining its own authority.

Practicing scientists prefer not to discuss this rather awkward conclusion which follows from their commitment to reductive materialism. In fact, this commitment is purely ideological.[28] They continue to treat science as a serious occupation. However, this lack of internal discussion and serious analysis has opened natural sciences to attacks from all kinds of postmodernist thinkers. They fall back on what amounts to an obscurantist position taking delight in the claim that our knowledge is inherently unreliable. Some of them suggest that science is just a social construct.[29] Of course, if no knowledge can be taken seriously, these postmodernist pronouncements should not be taken seriously as well. In fact, how can we take seriously anything?

Conclusion

From the scientific description of cosmogenesis one can make a good case that the appearance of humans was not a chance occurrence, but was prepared by a long, elaborate, and planned process. This conclusion comes as the fruit of the scientific enterprise. Although some of the greatest scientists have been religious believers, as a whole this enterprise has been and remains independent of religion and frequently even hostile to it. Today in the minds of many the goal of science is to disprove and to overturn religious belief. The fact that these efforts are increasingly seen as futile is remarkable. Although the most aggressive assaults on religion come from the so-called New Atheists (Richard Dawkins, Daniel Dennett, Christopher Hitchens, and their ilk) who keep on chanting the old mantra "science has proven that god does not exist," their arguments based on materialism have failed to persuade even some of their own colleagues. In that sense a recent book by the American philosopher Thomas Nagel aptly titled *Mind and Cosmos: Why the Materialist Neo-Darwinian Conception of Nature is Almost Certainly False* (2012) is most revealing. It demonstrates the materialism is self-contradictory in that sense that "evolutionary naturalism offers an explanation of our knowledge that is seriously inadequate, when applied to the knowledge-generating capacities that we take ourselves to have" and concludes that "conscious subjects and their mental lives are inescapable components of reality and consciousness is fundamental and irreducible and that comprehensibility of the universe is not a byproduct feature of contingent developments whose true explanation is given in terms that do not make reference to mind".

In fact, as I have tried to explain in the Introduction, the enterprise of scientific atheism is futile because it is contradictory. One cannot disprove Reason by using science, which assumes Reason as its foundation.

The parallels between the present narrative and that of first chapter of Genesis are, in my opinion, quite striking. Not being a theologian I do not insist on my reading of this sacred book, but on many occasions it does seem that Genesis gives seemingly absurd descriptions which turn out to be true. Thus light appears before the emergence of stars, which seemed to be absurd before the discovery of microwave background radiation; stars appear after the emergence of vegetation (indeed, the Earth's atmosphere became transparent only due to the vegetation: hence the stars and sun could not be seen before); animals of different kinds emerge explosively in contradiction to Darwin's theory but in agreement with the data (the Cambrian explosion). Most remarkably, although species change, new phyla do not emerge. It is also remarkable that the Hebrew word *barah* (to create) is used in Genesis I only three times—God (Elohim) creates Heaven and

Earth (which one may interpret as the spiritual and the material worlds), animals (the emotional sphere), and human beings. Only on these three occasions did fundamentally new things emerge. Nothing else, including primitive life, is said to be created and, indeed, it appears that on the level of cells there is no necessity to invoke any laws other than those which rule inanimate matter. This still leaves plenty of room for development and evolution.

The reader should not be left with the impression that the only thing science has done is to assure us of the validity of old truths. The picture of Creation painted by science contains many details to interest both the philosopher and the theologian. This picture gives the impression of a drama in which a superhuman wisdom is combined with a laborious effort. For the most part, wisdom reveals itself only gradually, over time. The origins of great undertakings frequently appear to be feeble, weak, and imperfect. Not every attempt is successful, as can be seen in the creation of intelligent life forms, which proceeded by experimenting with various candidates for the role now played by *homo sapiens*. There are disasters that seem to lay waste to the efforts of tens of millions of years of evolution, not to mention the enormous amount of suffering, killing, and death that has accompanied the process of life. Perhaps, we are both spectators and participants in a cosmic process of education and instruction? Could we believe that Nature as a whole undergoes the same kind of process? That would mean that Nature has a personality: wouldn't that be fascinating?

As is clear from these pages, our Universe is made for life. Is there life only on our planet or does life exist elsewhere? We do not know. However, there is more: the Universe is not just the home of life, even intelligent life:

it is also open to our rational inquiry. Although our minds have formed within the confines of this planet, they have been able to reach to the borders of the Universe, to the depths of atoms and atomic nuclei. Science—based on a belief in the rationality of Nature—has been tremendously successful. However, our success is not just the fruit of our effort; it is also a gift. If the Laws of Nature were a bit more complex than they are, we would not even be able to guess about their existence. Historically, the idea of natural laws came from the observation of repeated events, such as the motions of planets, the orderly change of seasons, the regularity of the tides. Without this approximate repetition we would not be able to verify the predictions of scientific theories. Although, as Heraclitus said, one cannot enter the same river twice, in many cases we can reasonably neglect the differences between similar events. This does not work for all processes; weather events and the turbulence of water, for example, are highly chaotic. However, according to the famous Kolmogorov-Arnold-Moser theorem of classical mechanics the quasiperiodic (almost repetitive) processes dominate over chaotic ones. So the regularity is a fundamental property of natural laws, at least on the level of macroscopic bodies. If these laws were different, so that nature was more chaotic, it would be difficult for us to study it. We would be in the situation we find ourselves in economics and history, where the very existence of laws is still debated. Let me quote Albert Einstein:

"You find it strange that I consider the comprehensibility of the world (to the extent that we are authorized to speak of such a comprehensibility) as a miracle or as an eternal mystery. Well, *a priori* one should expect a chaotic world which cannot be grasped by the mind in any way. One could (yes *one should*) expect the world to be subjected to law only to the extent that we order it through our intelligence. Ordering of this kind would be like the alphabetical ordering of the words of a language. By contrast, the kind of order created by Newton's theory of gravitation, for instance, is wholly different. Even if the axioms of the theory are proposed by man, the success of such a project presupposes a high degree of ordering of the objective world, and this could not be expected *a priori*. That is the "miracle" which is being constantly reinforced as our knowledge expands.

There lies the weakness of positivists and professional atheists who are elated because they feel that they have not only successfully rid the world of gods but "bared the miracles". Oddly enough, we must be satisfied to acknowledge the "miracle" without there being any legitimate way for us to approach it. I am forced to add that just to keep you from thinking that—weakened by age—I have fallen pray to the parsons".[30]

It seems that the world is made in a way so that we can acquire knowledge about it. Our ability to know and understand seems to be important to the One who created us.

Acknowledgements

I am very grateful to all those who helped me with this book, to Rev. Michael Meerson, Dr. Alexei Burov, Prof. Mikhail Arkadev, Dr. Elena Gorohovskaya who instructed me on the matters of biology, Prof. Andrew Schoffield, Larissa Pevear, Dr. David Shelton, His Eminence bishop Seraphim Sigrist and especially to my editor Penelope Burt.

Appendices

A. *The Anthropic Cosmological Principle*

In this section I will discuss aspects of the natural laws which allow and enable an uninterrupted increase in complexity, from elementary particles to DNA molecules, single-cell and multicellular organisms, and eventually to humans. It is now widely recognized that the existence of such an increase cannot be taken for granted, and that the very possibility of building more complex structures from simple ones imposes strict conditions on the natural laws.

To understand the factors that facilitated this cosmological process I will concentrate on a particular part of it which is relatively well understood, namely, the one that includes physical chemistry and the physics of condensed matter, with a direct relevance to biophysics. Although the so-called "Theory of Everything" has not yet been built, the "Theory of Everything Chemists Need (TECN)" has been known for quite some time, in fact, it is known to every physicist. For people familiar with physics, I would say that TECN is a theory which describes matter on low (subnuclear) energies and intermediate spatial scales (larger than nuclei, but smaller than planets). So it takes for granted the existence of stable atomic nuclei and does not discuss the effects of gravity. This theory can be formulated in a closed mathematical way; it includes the Schrödinger equation with the potential term depending on the position of nuclei and electrons interacting according to Coulomb's law and the spin-orbit interaction. The interested reader will find a more detailed description of TECN in Appendix B. In this theory, in addition to the values of the charges of the nuclei given by integers, there are only three "free" (i.e., not defined by the theory itself) parameters: 1) the so-called fine-structure constant α which characterizes the strength of the electromagnetic interaction; 2) the ratio of the mass of the neutron to proton mass (approximately equal to 1), and 3) the ratio of the mass of the proton to electron mass (approximately 1840).[31] It is striking that a theory with so small a number of free parameters can describe such an incredible diversity of the processes occurring in matter! Even more strikingly, the solutions of the equations involved describe the enormous amount of stable molecules of almost any degree of complexity involved in biochemical processes.

The problematic I am going to discuss below is directly related to the so-called Anthropic Principle (AP). In general terms, AP states that humans occupy a privileged position in the Universe. Originally this principle was put forward to give scientific justification to the old religious idea. However, in the course of heated philosophical and ideological debates, AP has lost any definitive form; and now groups with quite different metaphysical views adhere to its two main versions, called Strong and Weak AP. Depending on how one understands AP, it seems like either a perfect banality or a great intellectual achievement.

Strong AP is stated by Barrow and Tipler (1986, 21) as follows:

> *The Universe must have those properties which allow life to develop within it at some stage in its history.*

In what follows, however, I am going to be using a different formulation which is closer to the Discoverability Principle proposed by Alexei and Lev Burov (2016, 2016–2017), namely:

> *The Universe is designed not only to be inhabited by intelligent beings, but, much more, to be cosmically cognited by them.* (Moira and Eileithyia for Genesis, 2016)

Thus Strong AP uses science to support the idea that the Universe is designed with a special purpose.

Weak AP, in contrast, states that:

> *The features of physical laws beneficial to the appearance of Homo sapiens are the result of a self-selection process akin to Darwinian evolution: physical laws are life-friendly because otherwise there would be nobody to discover them.* (Cf. Barrow and Tipler 1986, 16, 19)

Normally, adherents of Weak AP assume an infinite or very large number of universes (the multiverse), each of them endowed with different laws of nature. Since the Universe we know is governed by the same laws of physics everywhere, the multiverse assumption is logically necessary to hold the Weak AP as an explanation of life-friendliness of the laws of nature.

Adherents of both forms of AP admit that the appearance of life in general and *Homo sapiens* in particular is conditional on the special properties of the laws of physics. The modern cosmology considers the Universe as an evolving entity in which more complex systems build upon and appear later than simpler ones. The bodies of intelligent creatures are the most complex material systems known, and they emerge at the latest stage of evolution. Therefore, the laws of physics must facilitate the uninterrupted process of emergence of ever more complex forms where every new step is *conditional* on the existence and stability of previous structures. Most discussions of AP revolve around this aspect of the problem. However, in my opinion, this unduly narrows the discussion; this is why I decided to use the rephrased formulation of Strong AP.

Opposite processes in cosmogenesis. In cosmogenesis one can clearly see two processes going on: the emergence of ever more complex forms and the degeneration, the irretrievable loss of some forms (rather than others), described by physicists as a process of increasing entropy. The newborn universe is simple and structureless and its entropy is low due to its incredible uniformity.[32] All material processes proceed due to this low entropy—and at the expense of its increase. However, entropy did not increase uniformly in all parts of our world. Every creation of complex structures is accompanied by a decrease in the local entropy for which it has to pay by the rise of entropy elsewhere. The most striking example of this is the emergence of life, which continuously creates new order locally, but increases disorder (entropy) in its environment. It is clear therefore that life is in some sense "unnatural" and very special; it runs contrary to the universal tendency towards destruction and decay and hence requires an explanation.

The fragility of complex structures. Complexity is a necessary condition for life, whether it is based on carbon or something else. This is due to the fact that biological processes require the handling of huge amounts of information and hence require facilities for its storage, protection, and processing. In the case of intelligent life, some of this information processing results in a knowledge of the world as a whole, and science is a part of it.

The more complex the structure, the more fragile it is, and the harder it is to find a place for it in the universe. To split a carbon nucleus requires a temperature of millions of degrees, which exists only inside stars; to detach electrons from the nuclei thousands of degrees are sufficient; to break up complex molecules one needs only hundreds of degrees (as in boiling water). With the emergence of protein-based life the complexity is several orders of magnitude higher than the complexity of structures of inanimate matter. The biosphere exists inside the geosphere and its existence is possible only in a narrow range of temperatures. In our solar system such conditions exist only on Earth and perhaps in the oceans of some of Jupiter's moons. Hence it is clear that the emergence of complex structures cannot be considered as a matter of course, that this cosmological process, like a stretched string, is always ready to break. Boris Pasternak wrote:

> You see, the course of the centuries is like a parable
> And can be stopped on the go.[33]

In earlier chapters I wrote about various interesting coincidences that prepared the conditions for the emergence of life. Recall, for instance, the emergence of carbon from triple collisions of helium nuclei. An analysis of this and similar miraculous "coincidences" is contained in Barrow and Tipler's seminal book, *The Anthropic Cosmological Principle* (1986).

In the present book I aim to explain how subtle is the balance of forces holding the complex system together and keeping it from disintegrating. More than that: I show how to express these apparent coincidences mathematically, in the form of mathematical restrictions on the physical constants that make up the laws of nature. Thus the conditions of possibility for life as we know it can be expressed as the conditions of existence for the solutions of this system of restrictions (equations and inequalities).

In his book *A Brief History of Time*, Stephen Hawking wrote: "The laws of science, as we know them at present, contain many fundamental numbers, like the size of the electric charge of the electron [the fine structure constant "*alpha*." See below—A .Tsv.] and the ratio of the masses of the proton and the electron.... The remarkable fact is that the values of these numbers seem to have been very finely adjusted to make possible the development of life" (Hawking 1988, 51).

In order to understand what Barrow, Tipler, and Hawking mean, we need to know what is meant by the fundamental numbers (in physics they are usually called dimensionless constants). I will give a brief explanation followed by a detailed example of the spectrum of the hydrogen atom.

The laws of physics are encoded in a set of mathematical relations between what can be called input data and output results. For every input, the output is determined by these relations (typically, by differential equations), which in turn are fixed by some general principles (see the discussion below), but not entirely. These relations frequently (not always!) contain numbers not determined by general considerations, and these are the ones that Hawking calls "fundamental."

AP as a resolution to a conflict. To appreciate the true meaning of the problem of fundamental constants one has to understand first why there are so few of them. This is because the form of the fundamental laws of physics is largely fixed by general principles mostly related to symmetry considerations.[34] It may seem like a paradox, but the fundamental laws possess a certain simplicity making them open to human discovery.[35] This simplicity carries with it a certain rigidity and inflexibility which, on the one hand, makes it easier for us to discover these general principles, but, on the other hand, sets significant restrictions on them which have nothing in common with life-friendliness, and thus could easily make the latter impossible.

The AP is frequently formulated as a problem of "fine-tuning" for life. From the perspective adopted in this book, such a formulation sends a very misleading message. As I will argue below, the constants of elegant laws cannot be "tuned" for life because there are too few of them to satisfy all of its requirements. The Strong AP reveals a conflict between the *general principles* according to which natural laws are organized and which are important for their discoverability, and the *life functions* these laws must facilitate. The general principles are largely fixed by requirements of symmetry and are distinguished by their terse elegance, leading to their ultimate *comprehensibility*. At the same time, the universe fashioned according to these principles must eventually provide room for complex beings capable of comprehending them. I will try to show that this conflict cannot be resolved by tuning (which does not exclude trial and error), but requires instead the mathematical structure of the laws to be specially chosen, thus requiring *intelligence.*

Dimensionless constants. For a better understanding of this important concept I will take as an example the simplest of all atoms—the hydrogen atom. Suppose we have a hot hydrogen gas; like any hot substance it glows and emits light. If we put the light beam emitted by the gas through a prism it will be split into a variety of colors (this is called spectral analysis). Each color is associated with an electromagnetic wave of a certain frequency— that is, the number of times the electromagnetic wave oscillates every second. It turns out that all frequencies emitted by hydrogen atoms can be represented by the formula (the Rydberg series):

$$\nu_{n,m} = (E_n - E_m)/h,$$

$$E_n = mc^2 [1 - \alpha^2/2n^2 + A(n)\, \alpha^4 + ...]$$

where ***mc²*** is Einstein's famous expression for the energy of the electron (this is the maximum energy that can be extracted from it, by completely destroying the electron), **h** is Planck's constant, alpha $\alpha = 1/137$ is the fine structure constant, ***n, m*** are integers, A (n) is also a certain number whose value is immaterial to us, and the ellipsises represent the higher members of the series in the "alpha²."

Now let me explain the meaning of these formulas. According to quantum mechanics, the energy of the stationary states of an electron in an atom can take only discrete values E_n corresponding to different values of the integer **n**. When an electron jumps from one "level" of energy to another, the atom emits light, the frequency of which is determined by the energy difference between the initial E_n and final E_m states as expressed by the formula (1). Equation (2) describes the possible values of the energy E_n of the electron in a hydrogen atom. It is written in the form of an expansion in the small parameter *"alpha"*. The first term of the expansion is simply the electron rest energy mc^2. The next term in the expansion, written in the explicit form, contains the square of the dimensionless constant *"alpha"*. All the other members of the series, symbolically marked by points, contain higher powers of this constant. Since electrons are held in atoms by the electromagnetic force, *"alpha"* characterizes the strength of the electromagnetic interactions. As can be seen from the formulas and the numerical value of *"alpha"*, these interactions are relatively weak. Indeed, the difference between the energy of an electron in an atom and a free electron is proportional to the *"alpha"* squared ~ 1/10,000. The similar formula for the energy levels in complex atoms contains as a parameter *"alpha" multiplied by Z*, where Z is the atomic number in the periodic table. For atoms with large numbers $Z \sim 1 / $ *"alpha"* the binding energy becomes comparable with the rest energy mc^2. This limits the size of the Periodic Table; atoms with numbers greater than about 130 are simply impossible.

This very *"alpha"* represents one of these dimensionless parameters that determine the structure of matter. It is dimensionless because it is not measured in meters or seconds; it is just *a number*. In the context of the "Theory of Everything Chemists Needs to Know" mentioned above, the numerical value of *"alpha"* has no justification. It appears as given and is approximately

1/137. In this theory, there are two other dimensionless parameters: the ratio of the mass of the proton to the mass of the neutron m_p/m_n and the mass ratio of the proton to electron mass m_p/m_e. They are also not determined by the theory itself.

General principles behind the laws of nature. As I have already mentioned, the general form of the laws in the "Theory of Everything" is fixed by certain fundamental principles that science, after much effort, has managed to discover. In making new hypotheses and constructing new theories scientists do not just blindly try whatever seems to fit the experimental data available at any given time. There are certain metaphysical assumptions they follow, consciously or unconsciously.[36]

These principles include, for example, general covariance (the laws must admit a formulation independent of the reference frame of the observer), the so-called CPT invariance, gauge invariance, and the principle of linear superposition of wave functions, etc. The first of these principles expresses the universality and timelessness of the laws of Nature. The second one states that the mathematical equations for the laws must be invariant under a simultaneous change of time direction, of parity (the interchange of right and left) and the replacement of particles by antiparticles. It would take too much time to explain the other principles. It turns out that these general principles largely fix the mathematical form of the natural laws.[37] The greatest experimental discoveries in physics, such as the discovery of electromagnetic radiation by Hertz in 1888 and the discovery of antiparticles in the 1930s, were predicated on the formulation of certain general principles.

What does all this have to do with life or, better yet, with us, human beings? As I have mentioned above, the simplicity of these fundamental principles stands in great contrast to the complexity of the world in general and life in particular. One cannot fail to notice that these principles seem to be quite abstract and devoid of human content. But they determine the entire variety of complex structures, including all the biochemical structures of our body, together with their functions. From these principles theorists derived the mathematical equations of the "Theory of Everything Chemists Need" (TECN) in their general form. Only the values of the dimensionless constants are left undetermined, at least for time being. Hence if the general principles remain unchanged (this assumption will be discussed later), then all the freedom to change the structure of matter, making it more or less stable, amounts to a change of these constants.

Why are the constants as they are? There are various suggestions, which I list below:

> 1. Since the laws of physics in general and the fundamental numbers in particular are such as to facilitate the emergence of life—and eventually of intelligent life capable of comprehending them, they reveal a purposeful design and hence point towards an intelligent Creator.

> 2. When we have a true Theory of Everything (ToE) that takes into account the dynamics of nuclear interactions and even gravity together with such factors as dark matter for which at present we have no explanation, we will learn that the values of the dimensionless constants are unambiguous. Such an answer implies that the universe cannot

be different from what it is and *homo sapiens* comes as part of a general package. This scenario opens up the possibility that intelligent life is just a byproduct of some general process.

3. The third answer is based on the fashionable Multiverse Theory (MT), which argues that our world is only one of an almost infinite number of universes, each with its own set of fundamental constants (or maybe even with different fundamental principles). It is alleged that the dimensionless constants in the "Theory of Everything, Chemists Need" are defined locally in each universe.

The scientific status of MT theory is quite shaky, but it is popular because of its ideological implications. MT exists in several forms. In its most extreme form it states that there are no general principles that fix the form of the laws in the entire Multiverse. In this form, it radically breaks with scientific tradition. It suggests that instead of being defined by general principles, the laws of physics are fixed by Darwinian selection, leading to Weak AP. Alexey and Lev Burov (2016) have shown that if this radical form of the multiverse (level IV Multiverse, Tegmark[38]) is true, our laws of nature would not be discoverable. Since they are discoverable to a great degree, this radical multiverse theory is refuted.

There is another form of MT, based on the yet unfinished string theory. This one admits general physical principles such as general covariance, gauge invariance, and the principles of quantum mechanics. In their attempts to remove the contradictions between the theory of gravity and quantum field theory, string theorists have arrived at a version of the Theory of Everything where the number of free parameters is almost infinite, much greater than in TECN. Each set of parameters is supposed to be implemented in its own universe, dictating its own laws of physics.

As I have said, in its extreme version the MT theory is not scientific. The quasi-scientific version of it is based on physical theories whose status is unclear. Their many aspects remain deeply controversial, and the theory has been challenged even by some of its founders, such as Prof. P. J. Steinhardt.[39] Below I will try to show that the Multiverse Theory is incapable of answering the question of why the world is not just fit for life, but is also comprehensible to intelligent beings.

Whichever version of MT one takes, the global picture is the same. Namely, in the vast majority of universes there are no complex structures, and hence no life, but we are fortunate to live in one (maybe the only one) where life is possible. Weak AP is frequently thought to mean that we would not be discussing the structure of natural laws if we were not here. At first glance this looks like an unbeatable argument but, as I am going to show, it is flawed. The first flaw is that it really misrepresents the situation. To discuss the laws of nature we must be aware of their existence. For most of its history, mankind has not even contemplated the idea of such laws.

Life-supporting laws may be self-selecting, but knowledge-supporting ones are not. We as intelligent creatures can be aware of laws of nature only in a world that is not just life-friendly, but knowledge-friendly. Since MT theory does not put any restrictions on the nature of physical laws, it does not preclude a situation in which the world is friendly for life just locally, perhaps even just for a single tiny planet in the vast Universe. On the other hand, to be knowledge-friendly is a very different matter. Such a world must be *globally comprehensible* for creatures endowed with a sufficient intellect. We could equally well live and even thrive in a world that would be utterly incomprehensible to us outside the boundaries of our own habitat. After all, there are many creatures on this Earth who are exactly in that situation and nevertheless surviving wonderfully. Many of them are a lot more ancient than human beings (horseshoe crabs, cockroaches, rats, just to name a few).

The extreme "everything is possible" MT has only one justification and, as was demonstrated above, it tries to achieve its purpose by slate of hand, substituting one argument for another. So there is no reason to discuss it further. One may argue that the quasi-scientific version of MT theory performs better, but the self-selecting argument fails there for the same reason.

Let us return therefore to our table of APs and consider statement 2, in which intelligent life is thought of as just a byproduct of some general structure of things. Is such a situation plausible? I will argue that it is not. This is because there appears to be a conflict between general abstract principles that lie at the foundations of physics and the *requirements of functionality* imposed by the very essence of life. These requirements emerge in multiple forms and can be mathematically formulated as restrictions on the dimensionless constants of physics. While looking through the pages of Barrow and Tipler's book, *The Anthropic Cosmological Principle*, one is struck by the fact that the same dimensionless constants appear in parts of the book that discuss completely different problems. There are pages that discuss the conditions for star formation, and pages that discuss the question of why the age of the universe must be several times greater than the time of biological evolution. Here they write about the number of elements in the periodic table (it is ∼ 1 / "alpha"), and there they write about the properties of water. All these topics are discussed because they are important for *performing certain functions* vital for the existence of life. A successful performance of each function imposes a certain condition on the dimensionless parameters. The conditions are many, but the number of these dimensionless parameters is few. In TECN, which restricts the consideration to the problem of stability of chemical structures, there are just three of them. Even if we add gravity and nuclear physics, there will still be less than ten. According to the quasi-scientific version of Multiverse Theory, these parameters are not fundamental, i.e., they depend on the infinite number of other parameters. If the Multiverse Theory is correct and the fundamental parameters are chosen randomly, then the ones of TECN are just random numbers. This amounts to saying that the problem of finding a world suitable for intelligent life is solved by trial and error.

Let us stop here for a moment and think about it. The Multiverse Theory tells us that the mathematical problem of finding solutions for life-supporting conditions was solved by trial and error. **However, not all mathematical problems can be solved by trial and error, and this includes the problem we are interested in.** The point is that while the stability of the material structures in our Universe is determined by only three dimensionless parameters, the number of conditions which they must satisfy is far greater than three! In fact, it is almost countless. Problems of this kind cannot be solved by trial and error because, as a rule, they have no solutions—unless the structure of the system is specifically designed for a solution to exist!

This point is extremely important and I would like to dwell on it a bit.[40] I ask the reader to be patient; making things absolutely clear may be dull, but at least the math is pretty simple.

I have said that the conditions for the existence of life (such as we know it) impose certain restrictions on the dimensionless parameters. To get an idea of how numerous they are, let us consider some examples. Imagine that you ate a piece of meat for lunch. Chemical reactions occurring in the body break down the meat into components, releasing energy which maintains life of the organism. The amount of released energy must be neither too little nor too great, because in the latter case the released energy could just break up the internal organs. (You cannot eat even a small piece of dynamite for lunch since its energetic content is too high.) A physicist can calculate the amount of energy released by a piece of meat. The resulting formula will contain the parameters "alpha", m_e / m_p, m_p / m_n (mass ratio of the electron and proton, proton and neutron) mentioned above. We can also calculate the amount of energy required to break the cells. So we obtain another expression, which, again, includes the same parameters. We now compare the two expressions. The amount necessary to fracture the cells must exceed the energy released after eating the meat. This represents one mathematical condition on the above parameters. Similar exercise can be performed with all the processes occurring in the human body, and each process will generate an additional condition on the same three parameters. We may consider conditions that the force produced by toothbrush will not knock out my teeth. Or conditions that any fall will not break my bones. Answers to all these questions can be expressed as mathematical inequalities containing just three unknown quantities, namely, the dimension-less constants.

I hope it is now clear that the number of mathematical conditions the parameters must fulfill is great, but the number of the parameters is small, only three. Let us now consider this immense heap of conditions as a system of equations (or rather inequalities), which one must fulfill in order to obtain the values of the parameters necessary for the existence of life, our life. When the number of equations exceeds the number of unknowns, mathematicians call it an *overdefined system*. For such systems, the existence of solutions is not guaranteed. However, we know that at least one solution exists—it is given to us in the form of the world as we know it![41]

Let us consider a simple example of a problem that cannot be solved by trial and error. Let's say we're playing roulette. On the drive there are 37 digits, including 0. The problem of obtaining the number 0 can be solved by just mindlessly spinning the roulette and 0 will appear sooner or later. However, since there is no number 38 on the drive, the problem of getting 38 digits cannot be solved in principle, even if one went on tapping for a billion years.

Now let us move to overdefined mathematical systems that also cannot be solved by trial and error. Here is a simple example. Consider a system of three equations for two unknowns, X and Y:

$$X + Y = 2, X - Y = 0, 2X + Y = 0$$

This system has no solutions. The first two equations have the solution $X = Y = 1$, but this solution does not satisfy the third equation. Imagine now that some bad mathematician tries to solve the system by mindlessly substituting different values of the variables X and Y. It is clear that since the system has no solution the trials will not provide any result even after billions of years. We have to refer the bad mathematician back to the elementary algebra course, which teaches that if the number of equations is larger than the number of unknowns, then as a rule, there are no solutions. However, this rule has exceptions, if the system is so cleverly designed that some of the equations represent a reformulation of others. Here's an example:

$$X + Y = 2, X - Y = 0, 2X + 2Y = 4$$

This system has the solution: $X = Y = 1$. It is possible because the last equation is essentially the same as the first (the left and the right sides are multiplied by 2), that is, the third equation is not a new one: it is simply equation number one in disguise!

Using more rigorous language we can say this: inside the set of *overdefined* systems of equations there exists an infinitesimally small specially constructed subset of solvable systems.

For an overdefined system to have a solution, it must somehow be specially designed or artfully constructed. In the case at hand, this art amounts to the correct choice of the original principles at the foundation of the laws of nature, providing solutions in the presence of an overcrowded system described above.

As I noted above, these fundamental principles include general covariance, CPT-invariance, gauge invariance, the principle of superposition of wave functions, etc. These principles are elegant and intellectually clear, which suggests a conscious decision.[42] The fact that in their very abstractness the fundamental principles do not explicitly contain human beings provides them in my eyes with a special elegance. God does not impose Himself, giving us a freedom to believe. I fully understand the feelings of the Artist, who challenged himself and set himself the task of building a universe of infinite richness and beauty using only a limited number of tools. This is the Anthropic Principle in its not banal formulation.

The argument for a conscious decision is so serious that it deserves a test. Suppose that space roulette implements not just different world constants, but also different principles of world order. Of course, one may ask with Steven Hawking "what is it that breathes fire into the equations" (Hawking 1988, 174), why the real world has to obey any rules prescribed by mathematical models.[43] However, let us assume that the roulette has this power to impose laws, but does it arbitrarily. For example, in one world there is the law of conservation of energy, and in another one energy is not conserved, one world has gauge invariance, the other does not, and so on.

So, for the sake of argument, we assume that anything is possible and there are no logical limitations on the universes, which are somehow generated in some kind of hyperspace. The only principle of selection that remains is our existence. We are simply in the world where we can live. Does this answer the riddle of world harmony?

Not really. The fact is, (this argument was put forward by my colleague, the physicist Alexey Burov) that we do not need just to explain our existence in the cosmos, we have to explain the fact that we are capable of comprehending it. "**Man** shall **not live by bread alone**, but by every word of God" (Luke 4:4). And the limits of our understanding, as I mentioned in the Introduction, are immeasurably wider than the limits of our environment. We are not just observers, as they call us in the theory of many universes, but *cosmic observers* or "comprehenders." And it is here that the real mystery lies, and the theory of many universes can say nothing about it. In this latter theory, our almost supernatural ability to comprehend just looks like a random fact. Isn't it too much for a random fact? According to this theory, we could quite happily live in a world that is incomprehensible to us globally, just as it is incomprehensible outside the narrow circle of their daily experience, not only for animals, but also for mankind in its pre-scientific era.

What makes the world comprehensible? It is not just its inherent logical structure (the laws of nature), but the compatibility of this structure with the human mind. The laws of nature might be so complex that we would not be able to grasp them, might never even guess that they existed. Imagine that the world had less regularity, so that the number of repeatable events—such as the change of seasons, the motion of the sun and planets, and so on—were smaller. In this more chaotic world, we would not even arrive at the concept of the laws of nature, though our bodies might adapt to such conditions. In fact, the regularity of events is determined by the particular structure of the physical laws (as evidenced by the famous theorem of Kolmogorov-Arnold-Moser, to which I refer the curious reader). Hence the comprehensibility of nature is conditional on the special structure of its laws.

Alexey Burov gives the following "impossibility formula":

> "According to the logic of the cosmological Darwinism, the proportion of cosmically-observed [i.e., comprehensible—A. Tsv.] worlds among all habitable worlds is zero. The condition of cosmic

observability of the universe is a very strong demand, additional to local observability, and, as such, strongly restricts the class of habitable universes. Impossible in this logic is not some kind of consciousness, but the one which is capable to observe and comprehend the universe as a whole, to be a cosmic observer."[44]

B. The Theory of Everything Chemists Need

In this Appendix I will demonstrate that the Weak Anthropic Principle is superfluous as far as chemistry in general and biochemistry in particular are concerned. The theoretical foundations of these scientific disciplines are well understood and are encoded in a system of differential equations which, as we will see, can be written on less than half a page. Every calculation carried out in these disciplines can be traced (usually through a series of approximations) to this foundational set. Hence it is legitimate to call the corresponding theory the Theory of Everything Chemists Need (TECN). This theory was briefly described in the previous section, here I will provide more technical details. These formulae contain only three dimensionless parameters whose values are not determined by TECN itself and hence can be considered as fitting parameters. The general form of TECN can be (and was) derived from general rational principles.

First, a brief introduction: TECN is not a theory of all matter; it concerns itself only with processes relevant for biology. Here to a certain approximation one can forget about gravity and nuclear energy. Gravity can be considered just as an external force, whose strength has a given value, which acts upon organisms; nuclei can be considered as the indestructible cores of atoms which have fixed mass and electric charge. With this caveat we can say that TECN can describe every physical process in the subnuclear realm.

TECN is a model in which matter consists of interacting atoms with indestructible nuclei. This is an approximate model which does not concern itself with such phenomena as radioactivity. All nuclei are positively charged, with an electric charge being an integer multiple of the electron charge e. An element whose nuclei have an electric charge $+Ze$ occupies place number Z in the Periodic Table of the Elements. Nuclei consist of electrically charged protons and electrically neutral neutrons.[45]

Isolated nuclei form atoms or molecules where positively charged nuclei bind an appropriate number of electrons (usually this number is equal to the atomic number Z of the given nucleus, so that the atom is electrically neutral). In condensed matter such as liquids or solids, electrons cannot be regarded as bound exclusively to the given nuclei. It is energetically advantageous for them to be shared between different nuclei, and this serves as one of the mechanisms of chemical bonding.

I will now proceed with the mathematical formulation of TECN. The fundamental equation describing the dynamics of a system of N nuclei with atomic numbers of Z_1, Z_2, ... Z_N and masses M_1, M_2, ... M_N and N_e electrons is described by the Schrödinger equation:

$$i\hbar\frac{\partial}{\partial t}\Psi\left(t;\vec{R}_1,...\vec{R}_{N_{nuc}};\vec{r}_1\sigma_1,...\vec{r}_{N_e}\sigma_{N_e}\right)=$$

$$=\hat{H}\Psi\left(t;\vec{R}_1,...\vec{R}_{N_{nuc}};\vec{r}_1\sigma_1,...\vec{r}_{N_e}\sigma_{N_e}\right)$$

Equation 1

where Ψ is the wave function, the knowledge of which enables one to calculate all observable quantities related to the given system. The wave functions contains as its arguments time t, the spatial coordinates of the nuclei R and the spatial coordinates of the electrons r together with their spins σ. Spin is the internal degree of freedom of an electron related to its angular and magnetic momentum. In other words, an electron resembles a spinning top and the projection of its angular momentum on a given coordinate axis is quantized and is equal to the product of Planck's constant and $\sigma = +\_1/2$. The wave function changes sign under permutation of any pair $(r_i, \sigma_i), (r_j, \sigma_j)$ (the Pauli principle). Since the nuclei are usually immobile, one does not need to take their spins into account.

The operator H acting on the wave function on the right hand side of the equation is called the Hamiltonian. Its form specifies the theory. The remarkable fact is that in the case of TECN the Hamiltonian can be written explicitly (all formulae below are written in SGS units):

$$H = -\sum_{j=1}^{N_p} \frac{\hbar^2}{2M_j} \frac{\partial^2}{\partial \vec{R}_j^2} - \sum_{k=1}^{N_e} \frac{\hbar^2}{2m_e} \frac{\partial^2}{\partial \vec{r}_k^2} + U_{Coulomb}\left(\{R\},\{r\}\right) + H_{spin\text{-}orbit},$$

$$U_{Coulomb}\left(\{R\},\{r\}\right) = \sum_{j<p}^{N_p} \frac{Z_j Z_p e^2}{\left|\vec{R}_j - \vec{R}_p\right|} + \sum_{j<p}^{N_e} \frac{e^2}{\left|\vec{r}_j - \vec{r}_p\right|} - \sum_{j=1}^{N_p}\sum_{k=1}^{N_e} \frac{Z_j e^2}{\left|\vec{R}_j - \vec{r}_k\right|},$$

$$H_{spin\text{-}orbit} = -i\frac{\hbar^2}{m_e^2 c^2} \sum_{j=1}^{N_e} \left(\frac{\vec{\partial}}{\partial \vec{r}_j} U_{Coulomb}\right) \cdot \left(\vec{S}_j \times \frac{\vec{\partial}}{\partial \vec{r}_j}\right).$$

Equation 2

It consists of the kinetic energy of the nuclei and the electrons (the first two terms in the first line), the electrostatic potential energy $U_{Coulomb}$, and the spin-orbit energy $H_{spin\text{-}orbit}$ which reflects the action of the electrostatic field of the nuclei on the spins of the electrons.

The last term constitutes a relativistic correction to the Hamiltonian. This form of TECN is applicable only to systems of those elements whose nuclei are not too heavy. (Such heavy elements as radium, uranium and beyond are radioactive and irrelevant for biology.)

It is already quite remarkable that complete information about all chemical processes can be packed into a single differential equation. More remarkable still is the fact that this equation contains only three dimensionless parameters. To see this we have to perform some simple operations. Specifically, I will rewrite the above formulae in dimensionless units, by introducing the following notations:

$$ r = \frac{\hbar^2}{e^2 m_e} x, \qquad E = \frac{e^4 m_e}{\hbar^2} \varepsilon $$

Now the distances are expressed not in centimeters, as in Equations 1 and 2, but in the unit of characteristic atomic size; energies E (recall that Hamiltonian has a dimension of energy) are expressed not in ergs, but in units of the ionization energy of the hydrogen atom. Now x and ε are just numbers. Rewritten in these units the Hamiltonian of TECN becomes *Equation 3*:

$$ h = -\sum_{j=1}^{N_p} \frac{m_e}{2M_j} \frac{\vec{\partial}^2}{\partial \vec{X}_j^2} - \sum_{k=1}^{N_e} \frac{1}{2} \frac{\vec{\partial}^2}{\partial \vec{x}_k^2} + u_{Coulomb}\left(\{X\},\{x\}\right) + \alpha^2 h_{spin\text{-}orbit}, $$

$$ u_{Coulomb}\left(\{X\},\{x\}\right) = \sum_{j<p}^{N_p} \frac{Z_j Z_p}{\left|\vec{X}_j - \vec{X}_p\right|} + \sum_{j<p}^{N_e} \frac{1}{\left|\vec{x}_j - \vec{x}_p\right|} - \sum_{j=1}^{N_p}\sum_{k=1}^{N_e} \frac{Z_j}{\left|\vec{X}_j - \vec{x}_k\right|}, $$

$$ h_{spin\text{-}orbit} = -i\sum_{j=1}^{N_e} \left(\frac{\vec{\partial}}{\partial \vec{x}_j} u_{Coulomb} \right) \cdot \left(\vec{S}_j \times \frac{\vec{\partial}}{\partial \vec{x}_j} \right). $$

This formula contains three dimensionless parameters:

$$\alpha = \frac{e^2}{\hbar c}, \qquad \beta = \frac{M_p}{m_e}, \qquad \gamma = \frac{M_n}{M_p}$$

—the so-called fine structure constant α, the ratio of the mass of the proton to the mass of electron m_e and the ratio of the mass of the neutron to the mass of the proton M_p. The latter two constants are encoded in the ratios of the electron mass to the masses of the nuclei:

$$M_j/m_e = (Z + N_n \gamma).$$

The approximate numerical values of these constants turn out to be $\alpha = 1/137$, $\beta = 1840$, $\gamma = 1.0013$.

If somebody played dice with our Universe, subjecting it to some kind of Darwinian process as the authors of the multiverse theory suggest, it is these three parameters which could be varied. All other parameters in *Equation 4*, such as the number of nuclei of a given kind, their charges Z, their number of neutrons, and the total number of electrons in the system are just integer numbers.

In practice the choice of these integers is very limited. The elements playing the most actively role in biochemistry are Hydrogen ($Z=1$), Carbon ($Z=6$), Nitrogen ($Z=7$), Oxygen ($Z=8$), Sodium ($Z–11$), Phosphorus ($Z=15$), Chlorine ($Z=17$), Potassium ($Z=19$), Calcium ($Z=20$), and Iron ($Z=26$). A tiny bacterium needs seventeen elements, and humans need twenty seven, including such exotic ones as molibdenum. As I have said, Equations 1 and 4 contain all the information about chemistry and even, in principle, all the information about the functioning of living cells (although in practice it might be difficult to find and interpret the solutions that would encapsulate this information). From this rather meager set of ten elements subject to the Schrödinger equation springs a plethora of biologically active molecules with innumerable functions. In order to be functional these molecules must have quite particular properties and there are too many of them to solve the problem by fine-tuning just three parameters. Of course, the parameters must have the right values, but this is not sufficient. The problem was solved not by throwing dice, but by an extremely clever choice of the basic principles of physics that determined the form of TECN.

C. Order and Chaos. The Early Universe as an Exceedingly Special State

As I mentioned in the main text, despite the lack of any definite forms in the early Universe it would be wrong to characterize its state as chaotic. To appreciate this fact we need to delve into the notions of chaos and entropy as the measure of it as they are understood in physics.

The notion of entropy is well defined only for macroscopically large systems—and the Universe is certainly one. In the statistical approach to such systems, one is not interested in such details as the behavior of every constituent particle. Instead we look only at *extensive* quantities averaged over macroscopic regions. By definition their magnitude scales with the number of particles in this region or with its volume such that the density of an extensive quantity remains constant. Such an approach can be illustrated by the example of air in a closed volume. At atmospheric pressure the number of molecules in one cubic centimeter is exceedingly large $\sim 10^{23}$. As I have said, we do not follow the movements of every molecule since these details do not affect such *extensive* quantities as pressure and total energy of the gas. The same pressure or total energy is realized for many different configurations of molecules. For instance, they will not be affected if we interchange their positions. The notion of entropy refers to the number of ways one can change the microscopic details of the system without changing the values of its extensive (that is macroscopic) properties. More precisely, the entropy is defined as a logarithm of this number. Let us consider a simple example. Imagine we have a collection of N cells with the weight of a certain mass positioned in each cell. Let our macroscopic state be characterized by the total mass of the system. This mass does not change if we permute the weights and there are $N! = 1x2x3x4x...$ xN such permutations. So the entropy of such a macroscopic state is $S = ln\ N!$ When N is much larger than 1 the factorial can be well approximated as $N! = e^{N\ ln\ N}$ ($e = 2.721828...$) so that $S = N\ ln\ N$. It is proportional to the total number of particles in the system and therefore it is indeed an extensive quantity.

According to Second Law of Thermodynamics, the entropy of a closed system always grows with time. So in the beginning our Universe had less entropy than it does now. Hence the number of ways to realize its initial state was much smaller than the corresponding number now. In that sense we say that it was more special in the beginning. To estimate the degree of this specialness we should look at the maximal entropy that will be achieved when all matter in the observable Universe collapses into black holes. A black hole is a perfectly featureless object, a perfect equalizer, the symbol of death itself. Hence it absorbs everything and nothing comes out of it; its entropy per a unit of mass is the maximum possible. The corresponding estimate has been made (see, for example the book by Roger Penrose *Fashion, Faith and Fantasy in the New Physics of the Universe*), and for the maximal entropy of the Universe it gives a stupendous number 10^{123}—meaning that the initial choice was approximately 3 to 10^{123} power. This is how special our Universe is (and may be even more). It is special, but in what way? Roger Penrose, who made this argument in his books (2005, 2016), explains that the sign of low entropy was the extreme homogeneity of the Big Bang. This fact has an immediate bearing on the status of Weak AP. To quote Chapter 3.10 of *Fashion...* : "A very striking thing about the low-entropy nature of our actual universe [it is related to its uniformity—A.T.] is that it is not just a local thing, operative only in our own neighborhood, but the basic structures—planets, stars, galaxies, galactic clusters—seem to proliferate in a roughly similar form... throughout the entire observable universe. ...Yet intelligent life on our own Earth needs only a tiny proportion of this volume of gravitational low-entropy. It is hard to see that our own lives depend upon such similar conditions holding in the Andromeda galaxy, for example, though perhaps some mild restrictions on it might be needed to prevent it from emitting anything dangerous to our existence. ... If we indeed simply requiring suitable conditions for the evolution of intelligent life here, than the figure of ten to the power 10^{124} that we appear to find for the improbability of the universe conditions that we actually seem to find ourselves in is ridiculously smaller than the much less modest figure needed for ourselves."

Longer Citations

To the Introduction

According to Loren Eiseley, the origin of modern science was rooted in:

> The sheer act of faith that the universe possessed order and could be interpreted by rational minds ... The philosophy of experimental science ... began its discoveries and made use of its method in the faith, not the knowledge, that it was dealing with a rational universe controlled by a Creator who did not act upon whim nor interfere with the forces He had set in operation. The experimental method succeeded beyond man's wildest dreams but the faith that brought it into being owes something to the Christian conception of the nature of God. It is surely one of the curious paradoxes of history that science, which professionally has little to do with faith, owes its origins to an act of faith that the universe can be rationally interpreted, and that science today is sustained by that assumption. (Eiseley 1961, 62).

To Day One

St. Augustine on the beginning of time. *The Confessions of St. Augustine*, Book XI. Trans. E. B. Pusey

> But if any excursive brain rove over the images of forepassed times, and wonder that Thou the God Almighty and All-creating and All-supporting, Maker of heaven and earth, didst for innumerable ages forbear from so great a work, before Thou wouldest make it; let him awake and consider, that he wonders at false conceits. For whence could innumerable ages pass by, which Thou madest not, Thou the Author and Creator of all ages? or what times should there be, which were not made by Thee? or how should they pass by, if they never were? Seeing then Thou art the Creator of all times, if any time was before Thou madest heaven and earth, why say they that Thou didst forego working? For that very time didst Thou make, nor could times pass by, before Thou madest those times. But if before heaven and earth there was no time, why is it demanded, what Thou then didst? For there was no "then," when there was no time.
>
> Nor dost Thou by time, precede time: else shouldest Thou not precede all times. But Thou precedest all things past, by the sublimity of an ever-present eternity; and surpassest all future because they are future, and when they come, they shall be past; but Thou art the Same, and Thy years fail not. Thy years neither come nor go; whereas ours both come and go, that they all may come. Thy years stand together, because they do stand; nor are departing thrust out by coming years, for they pass not away; but ours shall all be, when they shall no more be. Thy years

are one day; and Thy day is not daily, but To-day, seeing Thy To-day gives not place unto to-morrow, for neither doth it replace yesterday. Thy To-day, is Eternity; therefore didst Thou beget The Coeternal, to whom Thou saidst, This day have I begotten Thee. Thou hast made all things; and before all times Thou art: neither in any time was time not.

To Day Six

From a lecture by Karl Popper on the subject of evolution and consciousness (the first Darwin Lecture, delivered at Darwin College, Cambridge, November 8, 1977, Section 3. Huxley's Problem):

> The denial of the existence of mind is a view that has become very fashionable in our own time: mind is replaced by what is called "verbal behavior". Darwin lived to see the revival of this view in the nineteenth century. His close friend, Thomas Henry Huxley, proposed the thesis that animals, including men, are automata. Huxley did not deny the existence of conscious or subjective experiences, as do now some of his successors; but he denied that they can have any effect whatever on the machinery of the human or animal body, including the brain.

> "It may be assumed", Huxley writes, "...that molecular changes in the brain are the causes of all the states of consciousness... [But is] there any evidence that these states of consciousness may, conversely, cause ... molecular changes [in the brain] which give rise to muscular motion?" This is Huxley's problem. He answers it as follows: "I see no such evidence ... [Consciousness appears] to be related to the mechanism of ... [the] body simply as a col-

lateral product of its working ... [Consciousness appears] to be ... completely without any power of modifying [the] working [of the body, just] as the steam-whistle ... of a locomotive engine is without influence upon its machinery."

> FN (13) See T. H. Huxley, "On the hypothesis that animals are automata, and its history" (1874), chapter 5 of his Method and Results (London: Macmillan, 1893), pp. 239-40. While the passage quoted in the text refers to animals, Huxley follows it up, a few pages later, by saying "... to the best of my judgment, the argumentation which applies to brutes holds equally good of men; and, therefore, ... all states of consciousness in us, as in them, are immediately caused by molecular changes of the brain-substance. It seems to me that in men, as in brutes, there is no proof that any state of consciousness is the cause of change in the motion of the matter of the organism ... We are conscious automata..." (ibid., pp. 243-44). I have discussed these views of Huxley's in my paper "Some Remarks on Panpsychism and Epiphenomenalism", in *Dialectica*, volume 31, Nos 1-2, 1977, pp. 177-86, and in my contribution to *The Self and Its Brain* (see note 12 above).

Huxley puts his question sharply and clearly. He also answers it sharply and clearly. He says that the action of the body upon the mind is one-sided; there is no mutual interaction. He was a mechanist and a physical determinist; and this position necessitates his answer. The world of physics, of physical mechanisms, is causally closed. Thus a body cannot be influenced by states of consciousness. Animals, including men, must be automata, even if conscious ones.

Darwin's view of the matter was very different. In his book on *The Expression of the Emotions in Man and Animals* he had shown in great detail how the emotions of men and of animals can and do express themselves in muscular movements.

One direct reply of Darwin's to his friend Huxley, whom he greatly admired and loved, is most characteristic. A charming letter to Huxley written three weeks before Darwin's death, closes with a characteristic mixture of tenderness, irony, and wit: "... my dear old friend. I wish to God there were more automata in the world like you."

> FN 14 ([The Life and Letters of Charles Darwin, edited by his son Francis Darwin, John Murray, 1887 ... L. L., volume III, p. 358.)

In fact, no Darwinist should accept Huxley's one-sided action of body upon mind as the solution of what is called the mind-body problem. In his *Essay* of 1844, in his *Origin of Species*, and even more so in his much larger manuscript on Natural Selection, Darwin discussed the mental powers of animals and men; and he argued that these are a product of natural selection.

Now if that is so, then mental powers must help animals and men in the struggle for life, for physical survival. It follows from this that mental powers must be able to exert in their turn an important influence on the physical actions of animals and men. Animals and men could not, therefore, be automata in Huxley's sense. If subjective experiences, conscious states, exist—and Huxley admitted their existence—we should, according to Darwinism, look out for their use, for their adaptive function. As they are useful for living, they must have consequences in the physical world.

Thus the theory of natural selection constitutes a strong argument against Huxley's theory of the one-sided action of body on mind and for the mutual interaction of mind and body. Not only does the body act on the mind—for example, in perception, or in sickness—but our thoughts, our expectations, and our feelings may lead to useful actions in the physical world. If Huxley had been right, mind would be useless. But then, it could not have evolved, as it did, by natural selection.

My central thesis here is that the theory of natural selection provides a strong argument for the doctrine of *mutual interaction.* between mind and body or, perhaps better, between mental states and physical states.

Of course, I am very much aware of the fact that the doctrine of *mutual interaction* is utterly old-fashioned. Still, I propose to defend interaction, and old-fashioned dualism (except that I reject the existence of so-called "substances"); I even defend *pluralism*, since I hold that there are three (or perhaps more) interacting levels or regions or worlds: the world 1 of *physical* things, or events, or, states, or processes, including animal bodies and brains; the world 2 of *mental* states; and the world 3 that consists of the *products of the human mind*, especially of works of art and of scientific theories.[46]

Literature

Barrow, John D. & Frank J. Tipler. 1986. *The Anthropic Cosmological Principle.* New York: Oxford University Press.

Burov, Alexey and Lev. 2016. "Genesis of a Pythagorean Universe," in *Trick or Truth? The Mysterious Connection Between Physics and Mathematics,* FQXi, Ed. A. Aguirre et al. Springer, 157-169.

https://pythagoreanuniverse.com/files/Genesis-of-a-Pythagorean-Universe_Trick-or-Truth_Springer.pdf

Burov, Alexey and Lev. 2016–2017. "Moira and Eileithyia for Genesis," at "Wandering Towards a Goal" FQXi Essay Contest (2016-2017)

https://fqxi.org/community/forum/topic/2797

Carroll, Sean B. 2005. *Endless Forms Most Beautiful: The New Science of Evo Devo.and the Making of the Animal Kingdom.* New York: W. W. Norton.

Collins, Francis S. 2006. *The Language of God: A Scientist Presents Evidence for Belief.* New York: Free Press.

Denton, Michael 2017. *The Wonder of Water.* Discovery Institute Press, Seattle.

Eiseley, Loren. *Darwin's Century: Evolution and the Men Who Discovered It*. Garden City, NY: Anchor.

Hawking, Stephen. 1988. *A Brief History of Time: From the Big Bang to Black Holes*. New York: Bantam.

Jammer, Max. 2002. *Einstein and Religion: Physics and Theology*. Princeton, NJ: Princeton University Press. Kindle edition.

Koonin, Eugene. 2012. *The Logic of Chance: The Nature and Origin of Biological Evolution*. Upper Saddle River, NJ: Pearson Education.

Leslie, John, Kuhn, Robert Lawrence. 2013 *The Mystery of Existence: Why There is Anything at All*, John Wiley & Sons. Inc.

Nagel, Thomas. 2012. *Mind and Cosmos: Why the Materialist Neo-Darwinian Conception of Nature is Almost Certainly False*. New York: Oxford University Press.

Penrose, Roger. 2005. *The Road to Reality: A Complete Guide to the Laws of the Universe*. London: Vintage Books.

Penrose, Roger. 2016. *Fashion, Faith and Fantasy in the New Physics of the Universe*. Princeton, NJ / Oxford: Princeton University Press.

Plantinga, Alvin. 2007. "The Dawkins Confusion." http://www.christianitytoday.com/bc/2007/002/1.21.html

Schrödinger, Erwin. 1944. *What is Life? The Physical Aspect of the Living Cell*. Cambridge: Cambridge University Press.

Steiner, Mark. 1998. *The Applicability of Mathematics as a Philosophical Problem*. Cambridge, MA: Harvard University Press.

Whitehead, Alfred North. 1967 [1925]. *Science and the Modern World*. New York: Free Press.

Wilczek, Frank. 2015. *A Beautiful Question: Finding Nature's Deep Design*. New York: Penguin Press.

Recommended Reading

Barr, Stephen M. 2003. *Modern Physics and Ancient Faith.* Notre Dame, IN: University of Notre Dame Press.

Ferguson, Kitty. 2004 [1994]. *The Fire in the Equations: Science, Religion, and the Search for God.* West Conshohocken, PA: Templeton Foundation Press.

Gonzalez, Guillermo and Jay W. Richards. 2004. *The Privileged Planet: How Our Place in the Cosmos is Designed for Discovery.* Washington, DC: Regnery Publishing.

Plantinga, Alvin. 2011. *Where the Conflict Really Lies: Science, Religion, and Naturalism.* New York: Oxford University Press.

Tallis, Raymond. 2012. *Aping Mankind: Neuromania, Darwinitis and the Misrepresentation of Humanity.* Durham, England: Acumen Publishing.

Afterword

The Meeting of Logos & Nous in the Wisdom of Science

Michael Meerson

Pluralism in the Age of Information

No one browsing the Internet can doubt that we live in a pluralistic civilization, nor that a contemporary person is bombarded by cascades of facts, beliefs, philosophies, and mythologies, the secularist creed of human self-sufficiency being the most pervasive among them. The very process of industrial and technological progress generates this information explosion, thus making the pluralization of consciousness irreversible.

Science as the Leading Authority in Contemporary World

In our pluralist world, Science has come forth as the chief authority, partly because, paradoxically, it does not offer any final opinion on any subject. On the contrary, Science has offered a whole array of views, while also providing ways for their correction, refinement, and alternatives. Replacing Theology, Science, with its new methods, has tackled old subjects—namely, the human being and the universe—examining their innumerable aspects and dimensions. These scientific methods include critical analysis and comparison together with the empirical verification and systematization of data, ushering in ever-growing possibilities for the technological application of new discoveries. It keeps expanding the boundaries of what is known. Today any college student is taught to be apt at comprehensive searching for data on any subject, as well as weighing their various interpretations—just as a routine part of learning. These methods allow students to shape their own point of view, early on, on a variety of topics. Science has universalized these methods, applying them to all fields without exception. These fields include the one which previously had not required legitimation from Science—namely, Theology. In today's university, departments of religious studies are not circumscribed by any canon, be it the canon of the Holy Scripture, Sacred Tradition, or Eastern Patristics; be it the Papacy, Thomas Aquinas, Luther, or Protestant Neo-Orthodoxy. On the contrary, the contemporary university itself mercilessly scrutinizes these canons, deconstructing and evaluating them, only to return to them later, with a new research and analysis. But the same is true about other ideologies and doctrines—not only those based on religion, but also those opposed to it. Nothing can escape the scrutiny of scholarship or, if we use Kant's language, its "critique".

Logos and Nous as the Two Pillars of Science

From its inception Science rapidly grew in strength and influence. "Knowledge is power". Today no one disputes this motto of Francis Bacon, one of the founders of experimental science. Moreover, even the Western Churches, both the Roman Catholic and Protestant, giving rise to Academic Science which later abandoned its ecclesiastic cradle—even these Churches began to assess their doctrines and their credibility with the data and methods of scholarship.

But what are science and scholarship? Science is as old as human beings. After all, *Homo sapiens* from time immemorial has striven to know how nature was made, how it worked, and how one could utilize it. For Aristotle, the love of knowledge (*amor sciendi*), was the main virtue and passion of man. Anticipating contemporary Science, Philosophy, comprising such fields as Mathematics, Physics, and Astronomy, etc., arose in Ancient Greece. In his treatise, Professor Tsvelik, citing his great predecessors Einstein and Schrödinger, points out two premises of contemporary science. Science is possible because 1) there are universal laws governing events, and, 2) the human mind is capable of grasping these laws, thus being able to discover and to apply them.

These premises had originally been established by the Greek philosophy and science. Starting with its first thinkers, Anaximander, Heraclitus, Anaxagoras, through Plato, Aristotle and the Neo-Platonists, to the school of the Stoics, philosophy grounded itself upon these two principles: the universal Reason that kept Cosmos in order, and the human mind capable of knowing this Reason by partaking in it. Heraclitus called this universal Reason Λογος for its ability to communicate with man (Λογος from the verb Λεγω—"I speak"). Anaxagoras termed it Mind (νους), inasmuch as the human mind was able to understand it. One can trace the development of both terms through their interchangeable usage as the Greek thought developed. By the time of the Stoics their meaning was set, with the *Logos* assuming the stable meaning of the Divine Reason ruling the Universe. For the Greeks, though, *Logos* was not so much a god or a person, as an impersonal principle that kept the Cosmos in order.

A corresponding development can be traced throughout the Bible for the semantics of the Hebrew words *Binah* and *Khochmah*. *Binah* meant human understanding or intelligence, while *Khochmah* eventually came to mean the Divine Wisdom which can be imparted to the human being who possessed the faculty of Binah. *Binah* came to mean the understanding /epistemic intelligence, as the faculty enabling one to receive and internalize the Divine Wisdom (the *Khochmah*).[47] This meaning of *Khochmah* was finalized in its Greek translation as Sophia in the Septuagint, and passed on to the New Testament.[48]

For the Hebrews, the Wisdom Tradition played the role of Science, the same role as philosophy played for the Greeks. As Gerhard von Rad explained in his fundamental study, the Wisdom tradition originated "as a practical knowledge of the laws of life and of the world, based upon experience." It presumed a rational order underlying reality and promoted its understanding. Wisdom, as the discernment of reasonable order behind the flow of things and events, was an ancient way of scientific discovery. It expressed its noetic function in the form of numerous proverbs and maxims, first transmitted orally in a poetic fashion, and eventually forming the genre we know as Wisdom literature in the Bible. The Wisdom tradition anticipated later science, insofar as it "was open and never brought to conclusion," allowed antinomies without always trying to resolve them, and remained both "open to correction," and "capable of enlargement." The Wisdom tradition anticipated the method of Science in that it merged two different forms of acquiring knowledge—the "systematic (philosophical or theological)," and the "empirical and gnomic."[49]

This kind of science known as Wisdom was based on the Hebrew perception of the Deity as a Personal God, who had established a covenant with His people, Israel. After experiencing God's saving interventions in their history, the Hebrews developed the theology of God as the Creator of the world. Under His creative *"fiat"* (let there be) the universe gradually evolved in time, result-ing in the creation of "man, male and female, in God's own image and likeness" (Gen. 1: 26-27). The story, or rather the poetic hymn of creation which we read in the Bible's first chapter, was written in the metaphoric lan-guage that used the familiar imagery of the ancient Near Eastern culture. But through it, the biblical author(s) conveyed a unique theological vision, developed solely within Israel, with her distinctive experience of the Covenant. As presented in Wisdom literature, another principal gift of God to man was that of Wisdom which enabled humans to know God and His laws—first as moral rules to direct their lives, personal and social, and then, as the laws governing, or underlying the natural world. In Christianity, Hebrew Wisdom merged with the Greek understanding of Philosophy in the Person of the *Logos* incarnate who became human. The Christians, however, had recognized in the Greek impersonal cosmic Reason the second Person of God-the-Trinity, that is, as the co-creative Word. The *Logos* of the Christians, incar-nate in the God-Man Jesus Christ, reveals the ineffable God: "No one has ever seen God; the only begotten Son, who is in the bosom of the Father, he has made him known." (Jn. 1:18)

In the early Church, Athanasius of Alexandria, made this point with great passion and eloquence.

> The philosophers of Greeks say that the universe is a great body (comp. Plato, *Timaeus*, 30 &c); and rightly so,—he argued. If then, the Word (*Λογος*) of God is in the Universe, which is a body, and had united Himself with the whole and with its parts, what is there surprising or absurd if we say that He has united Himself with man also... For humanity, too, is an actual part of the whole. And as Mind (*Νους*), pervading man all through, is interpreted by a part of the body, I mean the tongue, without anyone saying, I suppose, that the essence of the mind is on that account lowered, so if the Word, pervading all things, has used a human instrument, this cannot appear unseemly.[50]

In Church theology, the *Logos* provides the underlying reasoning principle of Creation. *Logos* becomes human in Jesus Christ, who then assumes the function of a Teacher (Rabbi), instructing his followers as disciples. Being enlightened by *Logos* and his Wisdom, the man becomes capable of infinite cognition. The university system throughout the world has originated in these theological premises. It originally emerged as the institution of European Christian education, but eventually evolved into the universal form of the today's secular science.[51] Science is based on the conformity of the human mind with universal Reason as it was understood from the Middle Ages, through the Renaissance. This perception became even more firmly rooted in the Bible during the Reformation. This understanding was inherited by both the Protestant cultures and the modern times, even with their paradigm shift towards anthropocentrism.

Thus, modern science found its epistemological base on two pillars—using the familiar Greek terms—on *Logos*, Universal Reason, and on *Nous*, human reason, able to understand the workings of the *Logos*.

Recent studies have shown that Isaac Newton, who was chiefly responsible for science's current state and understanding, believed in this conformity of the human mind to the Divine. He synthesized two parallel traditions—Biblical faith in God as the Creator and Neo-Platonism—which viewed the Cosmos as reasonable and decodable. This synthesis, in turn, gave rise to a new method, that of mathematical physics. Newton followed the main Neo-Platonist postulate that Supreme Unity keeps the diversity of the world orderly and rational. He inferred that only a Supreme Rational Force could cause such a diverse yet elegant arrangement of the Sun, the planets and the comets within one harmonious system.[52] As the Russian historian of science I. S. Dmitriev notes, Newton proposed a model of the world which made its physical-mathematical content applicable beyond Newton's own context.[53] Marvelous as the implications of his model were for physical mechanics, they triggered the risk of interpreting both man and the universe in a mechanistic way.

By the time of Newton, Natural Philosophy had gradually split in two different directions progressively divorced from each other. Issued from Natural Philosophy, the natural sciences, in striving for objectivity, abandoned the human subject as a point of interest, and applied themselves to the study of physical "objective reality." In doing so, however, they began to neglect Science's fundamental principles: the *Logos* of objective reality and the Nous of the human subject's capacity to cognize it.

The Birth of Modern Philosophy as an Attempt to Understand and Explain Science

Modern philosophy, however, took another direction, by elaborating on these two principles, *Logos* and *Nous*, in order to bring their understanding to Science's contemporary state. Around that time and later, another paradigm emerged from the works of Descartes, Leibnitz and Kant. This paradigm constrained the mechanistic implications of Newton's model, and the purely objective direction that Natural Science followed after him. This was the paradigm of personal-centrism—a new philosophical interest in the cognizing subject. Descartes, with his *"cogito ergo sum"* (I think therefore I am), pronounced the conscious, thinking subject to be the very agent of scientific knowledge. With this discovery, modern philosophy emerged, as an effort to explain and understand the promising new reality of science. Science changed the place of man in the universe; humans now turned into the collective engineers of their own world. Four German thinkers—Kant, Fichte, Schelling, and Hegel—provided the philosophical foundation for modern science, by upgrading the usage of the concepts of *Logos* and *Nous*, as objective and subjective reasons. Kant and Fichte investigated the very mechanism of the human mind—*Nous*—behind the workings of Science. Schelling and Hegel brought to attention the working of the *Logos* in the universe and human history.

In his days, David Hume had also prepared the ground for a theory of scientific knowledge, by questioning Science's naïve claim that human cognition allegedly copies, or corresponds to, what the world is like.[54] He failed, however, to explain how Science worked. It was Immanuel Kant who endeavored to discover its mechanisms. Kant explored the discovery of Descartes' *"cogito"* as the basis of science. He worked out the epistemology that explained the nature of scientific, demonstrative knowledge, which had to possess two logical properties—necessity and universality. He agreed with Hume that empirical observation alone could not constitute science. Our everyday observation of sunrise and sunset does not suffice to prove that the Earth rotates around both the Sun and its own axis. Science allows us to draw logical conclusions from observations. It is thus rooted in the laws of the *Logos*. Kant reworked concepts inherited from the Greeks and the Stoics: namely the *Nous* and the *Logos*, examining both of them as human faculties. His notion of *Verstand/ Understanding* corresponds to the Greek *Nous* and the Hebrew *binah*. Luther in his Bible systematically rendered the Hebrew *binah* (understanding) as *Verstand*. Kant's Vernunft/*Reason* does not exactly correspond to *Logos*, but it does denote the supreme ability of our reason to unite all activities of understanding. "All our cognition starts from the senses, goes from there to the understanding, and ends with reason, beyond which there is nothing higher to be found in us. [Reason thus brings our work of cognition to] the highest unity of thinking."[55] As the One, in Neo-Platonism, keeps in unity all manifold ideas, so Reason, in Kant's philosophy, keeps in unity the rules by which Understanding operates. In Neo-Platonism, both the One and the cosmic sphere of Intellect lack personhood. In contrast, Kant, following Descartes, presents thinking as a personal activity. He regards the operation of both Reason and Understanding as the activity of the "I," that is, the cognizing Self. As forms of cognition, both Reason and Understanding are rooted in the "I." Kant called this "I" the "Transcendental Unity of Apperception."[56]

By "transcendental," he meant that all human subjects shared this property.

Although Reason abides within man, its origin is divine. According to Frederick Copleston's interpretation of Kant, what testifies to the existence of God is that the structure of all creation and of all thinking is reasonable. This testimony is stronger than the existence of contingent things.[57] Kant regarded Science as a strictly human activity, autonomous from both theology and metaphysics. Science deals with objective reality; God, in contrast, is neither present nor visible among objects. For Science, God is accessible only as an idea, existing within the realm of Reason.[58] Within Science, Kant distinguishes between the study of objects and the activity of the subject who studies them. He maintains that these two components are distinct and cannot be reduced to each other. After all, the subject studies his/her own subjectivity thereby, inevitably turning it into an object. An object, on the other hand, cannot study itself unless it becomes a subject. These, however, are people, not things. According to what Kant terms his Copernican Revolution in Epistemology, the universe rotates around our human self and mind, not the other way around. It is not that our cognition rotates around the universe. Summing up we may say that for Kant, Science is the method of the internalization of the world by the human mind, by *Nous*.

Kant's concept of the subject, fundamentally, parallels the terms of personalism.[59] Johann Gottlieb Fichte (1762–1814) further developed these personalist terms applying them to the process of cognition. In his book *Basis of the Entire Theory of Science (Wissenshaftlehre)*, Fichte expounded the concept of self (*Ich*, I) as the main working mechanism of science. It is the self-conscious human mind that keeps and organizes knowledge by bringing it into the unity of its self-consciousness. Fichte departs from the formal principle of identity "A is A" as the foundation of all knowledge, and demonstrates that this principle depends entirely on the intellectual activity of the I-subject.[60] It is the self-conscious subject who establishes that A is A. Every scientific discovery can be boiled down to this basic principle of identity: the expanding universe (A) is the same as the one 'I' have described in the series of equations on my pieces of paper (A') which describe the expanding universe (A). Though it may be described solely in the symbolic language of mathematics, my mental picture corresponds to the piece of physical reality in question. The basic principle of identity serves as the building block of every piece of scientific data—for example, this particular bacteria causes this particular disease. And this basic principle depends entirely upon the intellectual activity of some self, some "I." Fichte maintains that every thinker's awareness of thinking is what makes him/her a thinker. This awareness turns objects (that is, the *facts* of thought) into subjects, who engage in acts of thought. The subject is the "I," and the act is "I think."

Copleston elucidates Fichte's point with the following anecdote:

> Fichte once said to his students: "Gentlemen, think the wall." He then proceeded: "Gentlemen, think him who thought the wall." Clearly, we could proceed indefinitely in this fashion. "Gentlemen, think him who thought him who thought the wall," and so on. In other words, however hard we may try to objectify the self, that is, to turn it into an object of consciousness, there always remains an I or ego which transcends objectification and is itself the condition of all objectifiability and the

condition of the unity of consciousness. And it is this pure or transcendental ego which is the first principle of philosophy.[61]

Fichte's understanding can be traced in the whole course of modern western thought. At the beginning of the twentieth century, Edmund Husserl had come back to what Descartes discovered in the seventeenth, stating that "in reflection every *cogitatio* on being carried out takes the explicit form cogito."[62] Husserl further stipulates: "This visual ray [of *cogitation*] changes with every *cogito* shooting forth afresh with each new one as it comes, and disappearing with it. But the Ego remains self-identical."[63] Husserl thereby indicates the unique and crucial significance of every subject's "I think."

The Russian thinker Gustav Špet (Shpet) sums up this development. According to Shpet, most contemporary philosophers assume that the Self "is immediately present in every act of consciousness and experience."[64] John Searle, the contemporary American philosopher of mind, arrives at the same conclusion. He maintains that the ontology of the mind is subjective. The subject thinks in the first person, as "I think". Objects are thought about, they do not think. Any attempt to describe or research one's mind objectively, in the third person, then reduces the "I" to an "it", the subject to an object. Subjectivity cannot be reduced to an object of study alone.[65]

This philosophical discovery that mind is personal and subjective corresponds to Christian Theology. Starting with John's Gospel, Christ the incarnate Word of God, is identified with the Logos of the Greek thought. But this means that the Logos is personal. Just as our human thinking is always done by a subject/person/someone, by *myself*,—so the universal Reason must also be personal, not merely an impersonal principle. Christ's divine "I" is the "I" of the Trinity, paradoxically, for all three Persons of the Trinity. Thus, for Christian Theology.[66]

Likewise, in philosophy. As Fichte maintains, inasmuch as science is an activity of the mind, the self is the first principle of all knowledge. Some contemporary philosophy further identifies the subject as an agent; as John Macmurray states, "The Self that reflects and the Self that acts is the same self, action and thought are contrasted modes of its activity."[67]

This fundamental distinction, between the subject of thinking and the object of thought, has been blurred in the subsequent development of the so-called "Scientific Worldview"—or rather, of the idolatry of Scientism. Let us, however, return to Kant. After all, in a way, he was the one who had triggered the emergence of this "Scientific Worldview." Half a century before Laplace, Kant, a scientist in his own right, set forth the Nebular Hypothesis of the birth of the Solar System.[68] He did it in his early writings, mainly scientific, primarily in his *General Natural History and Theory of the Heavens* (1755).[69] There, Kant worked out his theory of global evolution, describing the nature of our universe as emerging and evolving according to objective and immutable laws.[70] He considered the regularity of this evolution symptomatic:

"Nature was created by a reasonable Being according to His ends and purposes."[71] Kant's early works laid the foundation of scientific Cosmogony, introducing the notion of development into natural sciences. Unlike the French Enlightenment thinkers, Kant emphasized that this development could not be explained by any mechanistic model alone. Biology cannot be reduced to mechanistic laws. They cannot explain the emergence of even a single blade of grass or a caterpillar.[72]

After Kant, Hegel presented the evolution of the World as permeated with thought. He set forth a vision of the unified development of Cosmos, Earth, Nature, and, finally, Human Culture. This development, Hegel maintained, was that of one absolute Idea. Copleston equates Hegel's Eternal Idea with the *Logos*.[73] Displaying this Eternal Idea in action, Hegel's system consisted of three parts. The first part, the Science of Logic, laid out the metaphysical rules by which the Logos operates. In the second part, Hegel traced the action of the Logos in the development of Nature and the Cosmos. In the third part, he develops a philosophy of Mind/Spirit (*Geist*), tracing the presence of the Logos in the evolution of human civilization.

Hegel used Kant's terms for Understanding (*Verstand*) and Reason (*Vernunft*), firmly placing them, however, within the realm of the Eternal Idea. Kant had set the notion of the Thing-in-Itself as both an aide and an obstacle to our cognition, insisting that our reason not only discovers the world but also screens its true essence from us. He insists that the "Thing-in-Itself" is unknowable. Hegel, on the contrary, asserts that our reason can discover how the world is "in itself," because our reason participates in the Mind who governs the world. He refers to the Greek Anaxagoras who was the first to announce this truth.[74] Our reason is equipped to cognize the world, because "the very same reason that is in us, is in the world itself as its own defining principle."[75]

For Kant, Science is a *method* by which the human mind internalizes the world. For Hegel, Science is also a *process* which the human mind uses to internalize both History and the World. He describes this process as dialectics. At first Hegel develops his own dialectic method by critiquing Plato's dialectics. In his dialogues, Plato displayed the movement of two opposing views toward a unified vision that preserved the content of both views in a synthesis. In Plato's dialogues, Socrates, the champion of dialectics, usually wins. But his victory entails a synthesis of opposing opinions. Socrates considers this resolution and unification of opposite views to be the essence of Science.[76] Socrates, who calls himself "a lover of learning," and "a man of Science,"[77] defines scientific knowledge as an ability to bring "a dispersed plurality under a single form, seeing it all together."[78]

Hegel claims that, while developing this method, he raised it "to the level of a genuine science."[79] Reality itself is so rich and contradictory that reason generates contradictions while trying to grasp it. In the process, Reason overcomes these contradictions by taking them to a new level, and eventually, to Supreme Unity.[80] This unity, however, is also a moment in the process of infinite becoming. This process keeps generating ever new opposites, to be overcome in ever new and ever higher unities. No external cause drives this process; it is purely mental. Understanding (*Verstand*) works by abstracting a piece of reality as a concept, thereby producing a one-sided abstraction. This one-sided abstraction from reality evokes a concept that contradicts it and is, in turn, also a one-sided abstraction. The first concept generates its opposite out of the inner necessity known as Reason (the Vernunft) in its capacity to encompass totality. By the very act of generating its opposite the first concept both allows for its own limitation and sublates it—to use the English rendering of Hegel's term *aufheben*.[81] The second contradiction, in turn, also admits its own limitation, thereby striving to overcome it by a new concept that would reconcile the two opposites. The new synthetic concept, in its turn, becomes a new thesis. On every level, each thesis is immanently conditioned to generate its own antithesis to be reconciled in a new synthesis—ever approaching and never reaching the Supreme Unity. As this Supreme Unity, Absolute Idea, or Logos, permeating all of reality, drives it on by this dialectic process. One can detect this process in Nature, history, and human cognition. In Hegel's mind, Dialectics is "the principle through which alone *immanent coherence and necessity* enters into the content of science."[82]

Scholars still argue whether Hegel meant something transcendent or immanent by Logos. Did he think that Logos was an intellectual faculty of mankind or a divine force manifesting itself in human cultural activity?[83] In either case, Hegel developed the concepts of *Nous* and *Logos* which he had inherited from the Stoics and the Neo-Platonists, including their Christian adaptation.

The very ambiguity of Hegel's thought comes from his dependence on these traditions, both philosophical and theological—whose true interpreter, he claimed to be. He employed the language of Christian theology, including its biblical Personalism. Hegel, however, dissolved this language in the Neo-Platonic system, its Logos and the One being equally impersonal. Nonetheless, Hegel represented, heralded, and imprinted the Modern Age, with his own kind of personalism. His Mind/Spirit (Geist) possessed both freedom and self-consciousness. The Mind/Spirit gains its own self-consciousness through the personal self-consciousness of man. This Hegelian idea itself stems from the Christian belief that the Logos became Man as Jesus Christ.

After Hegel, Schelling specified Science's contribution to Religious Mind as follows: Science presented Christian Revelation as the ultimate point in the mankind's religious development.[84] All four thinkers—Kant, Fichte, Schelling, and Hegel—came from the tradition which has engaged in a rational exploration of the Scriptures. As the springboard for their thought, this exegetical tradition helped them to understand the new phenomenon of science. By the end of the nineteenth century, the neo-Kantian Hermann Cohen aptly characterized Kant's Critique as a "critique of pure science, that is, a theory of science."[85] The four thinkers' philosophy reflected a paradigm shift in modern culture, now centered on man equipped with scientific method. Kant and Fichte concentrated on the workings of the human mind as a lofty personal faculty and activity, while Schelling and Hegel showed the working of Universal Reason in Nature, human history and culture. Thus German Classical thought accounting for modern science revisited, on a more advanced level, the Greek concepts of Logos and Nous as the two pillars of ancient Science.

The philosophy of German Classical Idealism interpreted the phenomenon of Science as a whole—not just as a multiplicity of branches. Schelling, for example, stressed the significance of the University as the place where philosophy would find a unified vision of these branches—not merely as a place to study multiple disciplines. After the Reformation, the multiplication of religious doctrines called for philosophy's unifying hand. The same happened with Science's coming of age—which has also required its own unified philosophical understanding.

Meanwhile, Science itself was growing both more powerful and more fragmented. As Sergey Bulgakov maintained, from the mid-nineteenth century onwards the positivist thinkers, such as Auguste Comte and Herbert Spencer, attempted to overcome this fragmentation by creating a taxonomy of the sciences.[86] But their attempts to create a synthesis of sciences into one unified scientific world-view quickly grew outdated after each new scientific development. Unlike Philosophy, Science does not aim to generalize. In Bulgakov's view,

> scientific knowledge is not and cannot be summed up in any kind of synthesis: *increasing specialization is a law of scientific progress*. A *scientific*, as opposed to a *philosophical*, synthesis of the sciences into Science is a utopia, for science has no way out of the empirical world, where all is multiplicity. We must not forget that sciences create their own objects, set up their own problems, and determine their own methods. There can thus be no single scientific picture of the world, nor can there be a synthetic scientific worldview. Each science yields its own picture of the world; it creates a reality of its own, which may or may not resemble the reality of another.[87] (Italics mine—MM)

Are the different sciences then completely disconnected? Following the neo-Kantians, Bulgakov argues that "sciences are connected among themselves by their formal aspects, their methodism, the logical techniques of concept formation, rather than by their content."[88] In other words, it is logical structure—*Logos*—that underlies the reality of both life and Science, and the human mind—*Nous*—that grasps it. What ensures and explains the unity of Science is a philosophical claim—the Anthropic principle as Alexey Tsvelik explicates it. The sciences, Bulgakov maintains, "are united in the oneness of their (transcendental) subject—man as universal humanity—and in their substratum—the single all-penetrating and all-creating life, which generates them from its womb, from mysterious and immeasurable depths."[89] It does not mean that physics and microbiology provide us with incompatible models of the world. Each one creates its own model. They need philosophy to bridge these models and show their compatibility with each other, precisely by discovering in each one the presence of *Logos* and the working of *Nous*.

The Emergence of the Scientific Myth

Modern science grew from Natural Philosophy which gradually split into two different fields—Science and the Humanities, progressively divorced from each other. In their strivings for objectivity the Natural sciences expelled the subject. In the method of natural sciences, the notion of the observer was completely abstract. Science regarded *everything* as an object, and, in studying it, tended to forget about the subject who is doing the studying. In the words of Tsvelik, this is a difficulty Science has never been able to resolve.

Relying on Science and its preoccupation with objectivity, the creators of the Positivist scientific worldview, also ignored the thinking Self, the Subject of thought. Newton's formulation of the laws of Classical mechanic not only brought about the industrial revolution of the eighteenth and the nineteenth centuries, but has also provided a seemingly scientific basis for constructing a new mechanistic model of the world. The materialist thinkers of the eighteenth-century French Enlightenment created a hypothetic world-view which took the shape of a new myth. At the end of the nineteenth century, Ernst Mach, the Austrian physicist and historian of science, described this process:

> The French Encyclopedists of the 18th century thought that they were not far from a definitive explanation of the world based solely on physical and mechanical principles: Laplace even imagined a mind that would have the power to predict the course of nature for all eternity, once the masses of all bodies, their positions, and their initial velocities were given. In the 18th century the gleeful optimism about the capacity of the new physico-mechanical ideas was pardonable. It was even a refreshing, noble, and lofty spectacle; and we can deeply sympathize with this expression of intellectual rapture, so rare in the annals of history. But now, more than a century later, when our judgement has become more sober, the Encyclopedists' conception of the world appears to us as a mechanical mythology, not all that different from the animistic mythology of the ancient religions. Both these views contain the false and fantastical exaggerations typical of a partial perception.[90]

The Discovery of Matter as a Living Body

In fact, by the mid-nineteenth century, with the advancement of biology, Laplace's mechanistic model of the Universe became outdated, just as the mechanistic myth of the Encyclopedists did. Science discovered the world as a living body and the matter itself as a productive, developing reality driven by its inner dynamism. To be sure, this discovery did not occur without philosophy's influence. After Kant, with his hypothesis of global evolution, and Hegel, with his teachings on the Idea-Logos evolving throughout history, the understanding of the world as a static entity rapidly began to wane. Everything that existed came to be perceived as growing and evolving toward greater sophistication. The philosophy of Hegel, "that apostle of History,"[91] triggered the advance of historical studies in the modern age. It introduced the principle of historicity into various fields of Science. As Sergei Bulgakov wrote, Schelling, in his own words, "broke through to the free and open field of objective science," introducing "the understanding of nature as a living, growing organism."[92] Schelling actually designed a blueprint for future theories of evolution, including dialectical materialism. According to him, nature, being a productive entity, is more than an object. As the subject of all possible productions, Nature is impregnated with reason.[93]

Schelling's reason, however, differs from the transcendent *Logos* of Christian revelation. The living God—who, in the words of Pascal is the "God of Abraham, Isaac, and Jacob, not the god of scientists and philosophers—obviously remained in the Church, even divided, but He was now locked up within it. Philosophy informed by Science left God outside. On the other hand, the churches, albeit numerous now, each with its own dogmatic system, each contradicting another, also attempted to keep Science and its philosophical interpretation under control and contained within their own walls. After the Enlightenment, however, the Church could no longer do that.

Kant, among others, sought to provide Science with "a safe zone." In the subjective world, Kant still vouchsafed God a place in the realm of morality. As for the world of objects, the Philosopher entirely removed God from it. This view also minimized the creative role of the Logos. Initially following Kant, Arthur Schopenhauer eventually dismissed the relevance of the Logos altogether, replacing it with a blind universal Will, interpreted as the impersonal "libido" of sexual reproduction, which animated and ruled the world. Scholars started taking philosophical systems apart using their separate fragments, torn from their original context, to suit their own needs and disciplines. Freud, for example, developed his own theory of the Subconscious based on Schopenhauer's theory.

In the nineteenth century, Science, excited by its own achievements, was considered to be the main engine of human progress. Science, however, clashed with the fundamentalist reading of the Bible by the Protestants. On the other hand, it also collided with the mechanistic Aristotelian metaphysics of the Roman Catholic Church. Ironically, around the same time the First Vatican Council (1869–70) canonized the Thomist scholastic theology based on Aristotle. This canonization de facto turned the Catholic Church into an obstacle to scientific advancement—even without intending to antagonize it. The Thomist doctrine failed to develop a new anthropology fit for man's new awareness of his role and place in the world.

The New Myth of the World and the Human Being Produced by Self-Engendering Matter

Opposing the ossified theology discussed above, a new atheistic myth made its appearance—the myth of the eternal, uncreated matter, self-engendering and self-correcting.

This myth revived the pagan myth of an eternal and uncreated world. This new version of the myth was adjusted to the new scientific age. It generated a new ethics and ontology, devoid of personhood. Ludwig Büchner (1824–1899) heralded this world-view, in his book *Kraft und Stoff/ Force and Matter: An Essay on the Natural World Order with Teaching on Morality Based upon it*. A physiologist by training, Büchner patterned organic life after the mechanistic model of a steam engine. Just as a steam engine triggers motion, so a certain living force in the body generates a type of behavior known as a combination of "Soul and Consciousness." For Büchner this living force is rooted in matter. It turned out that the deterministic world-view of Laplace was compatible with physiological determinism of Büchner.

Yet Büchner was not consistent about his materialistic explanation of the mind, sometimes acknowledging it as something autonomous. An atheist and a materialist, Büchner rejected both the Reason of the Idealists and the God of the Christians. He regarded them both as obstacles to scientific progress. In Russia, this new materialist myth became so popular that the Russian translation of Büchner's book was reprinted over 20 times.

But it turned out to be popular beyond Russia and Germany. New conceptual models emerged aiming to replace the Creation narrative of the Bible. Among them, Herbert Spencer (1820–1903) introduced a new doctrine which became popular overnight: the theory of an all-compassing development of matter and human society. This doctrine is popular even today, albeit in our collective unconscious mind. The empirical data Spencer used and interpreted quickly became outdated. His theoretical inferences were soon dismissed as philosophically unsound. His name has fallen into oblivion. Yet, despite all this, the mythology he fashioned has endured even up to the present. After all, it relied on the popular belief in scientific progress and mankind's self-sufficiency.[94] Spencer worked out an all-encompassing doctrine of Natural Evolution. According to Spencer, the physical world evolves into animal organisms, which in turn evolve into the human mind, which in turn enables human civilization to evolve into some vaguely advanced and sophisticated organization. All this progress is triggered by a force of some indefinite sort—which Spencer failed to explain.

Initially Spencer had shared Kant's phenomenology and theory of evolution. In it, as we recall, God the Creator remains outside the world of objects, and thereby, also outside the domain of scientific cognition. Spencer also used the Neo-platonic theory of Unity and Multiplicity to interpret the relationship between Science and Philosophy. In his view, common folk knowledge is un-unified—the lowest kind of knowledge; Science partly unifies it, but only philosophy makes it whole.[95] Spencer claimed that he himself reached this wholeness by generalizing data from various branches of science. As a positivist (unlike Kant and Fichte), Spencer dismissed the centrality of the personal cognizing subject. Ignoring both Universal Reason and God the Creator, he relied on the hypothesis of the Universe's blind evolution, based on the First Law of Thermodynamics. According to him, all structures in the universe develop "from a state of relatively indefinite, incoherent homogeneity to a state of relatively definite, coherent heterogeneity."[96] This process of gradual development towards more order and complexity has been at work on all levels, from the formation of stars and galaxies, to the biological evolution, to the advancement of human society and its political and economic order.[97]

According to Tsvelik, this attempt to deduce the process of self-organization from the First Law should be considered as a complete failure. The First Law is just another formulation of the law of the conservation of energy. To justify the emergence of new forms by a process of conservation is completely illogical. Nothing creative follows from the conservation laws and, certainly, something else is needed to propel any self-organization.[98]

While relying on the First Law of Thermodynamics, Spencer overlooked the Second Law, that of Entropy—which annuls Spencer's whole philosophical edifice. The Second Law implies that, left to its own devices, the world would regress to chaos and disorder, that is, the state of thermodynamic equilibrium which physicists call the "heat death." Instead, to ensure order, Spencer introduces a Deus ex machina metaphysical principle, which, just like Büchner, he terms "Force." Spencer calls this Force "the ultimate of ultimates." It grounds all our basic concepts—Space, Time, Matter, and Motion. This Force is not material, and it cannot be reduced to energy either. Spencer, in fact endows this metaphysical principle with reason: he considers this Force to be "the basis of Science." With this Force as its foundation Science must therefore take this Force for granted. In Copleston's words, Spencer's "Force" "cannot be established by science."[99] Spencer spoke as an apostle of science; yet, he failed to explain where thought itself, and the science engendered by it, came from.

Some erroneously believe that Spencer based his Evolution myth on Darwin's theory of evolution. In fact, after Kant, Schelling, and Hegel, this idea was in the air. In his autobiography, Darwin wrote that he had inherited this idea from his grandfather, Erasmus Darwin.[100] Spencer and Darwin came from the same intellectual milieu. Spencer's father, William George Spencer, a dissenter preacher and free-lanced philosopher, served as the Secretary for the Derby Philosophical Society, founded by Darwin's grandfather. Whatever education in science, physics, mathematics, and Latin he had, Herbert Spencer received it from his father and uncle, and from other members of the Derby Society.

Spencer's idea of the progressive evolution of life might have been influenced by Robert Chambers' *Vestiges of the Natural History of Creation* (1844). Chambers had combined the idea of the divine creation with Kant's and Laplace's Nebular Hypothesis of the development of solar system and with Lamarck's hypothesis of biological evolution. Spencer, also following Lamarck, formulated his theory of evolution two years prior to the publication of Darwin's *The Origin of Species* (1859).[101] After reading Darwin's book, Spencer summed it up in his own words—it was all about the "survival of the fittest." Darwin appreciated the description, including it in a subsequent edition of

The Origin. Natural Selection

In cultural memory, however, the theory of Evolution is associated not with Spencer and Lamarck, but with Charles Darwin. This theory entails an anthropogenesis that completely abolishes the perception of the human being as the subject of knowledge. Rather, it presents man as an object blindly produced by nature and submerged in it. Darwin knew Spencer and appreciated him as a thinker:

> Herbert Spencer's conversation seemed to me very interesting... After reading any of his books, I generally feel enthusiastic admiration for his transcendent talents, and have often wondered

whether in the distant future he would rank with such great men as Descartes, Leibnitz, *etc.*, about whom, however, I know very little.[102]

Yet Darwin points out that he would never consider Spencer a scientist:

> Nevertheless, I am not conscious of having profited in my own work by Spencer's writings. His deductive manner of treating every subject is wholly opposed to my frame of mind. His conclusions never convinced me: and over and over again I have said to myself, after reading one of his discussions,—"Here would be a fine subject for half-a-dozen years' work." His fundamental generalizations (which have been compared in importance by some persons with Newton's laws!)—which I daresay may be very valuable under a philosophical point of view, are of such a nature that they do not seem to me to be of any strictly scientific use. They partake more of the nature of definitions than of laws of nature. They do not aid one in predicting what will happen in any particular case.[103]

While Spencer deduced his universal system of development from the First Law of thermodynamics, Darwin limited his study to biological evolution alone. The phenomenon of man was lost in the vast expanses of biology; hence so were lost both *Nous* and *Logos*. As Darwin himself stated, he neither knew philosophy, nor was interested in it. Accordingly, he ignored all the concerns and considerations of philosophy. A systematizing collector of empirical data, Darwin nonetheless could not unify them into a coherent theory. As he himself admitted, he borrowed his theory from Thomas Malthus's *An Essay on the Principle of Population* (1798). Malthus's book influenced Darwin's notion of a mechanism of evolution that he called Natural Selection. Darwin wrote:

> In October 1838, that is, fifteen months after I have begun my systematic enquiry, I happened to read for amusement Malthus on Population, and being well prepared to appreciate the struggle for existence which everywhere goes on from long-continued observation of the habits of animals and plants, it at once struck me that under these circumstances favorable variations would tend to be preserved, and unfavorable ones to be destroyed. The result of this would be the formation of new species. *Here, then, I had at last got a theory by which to work.*[104]

Theorist or not, as a biologist Darwin acknowledged the complexity and organization present in every living organism. His theory of evolution plainly speaks of order as its feature. Since Aristotle, complexity and organization have belonged to the domain of reason. Darwin tacitly allowed their presence in life processes by comparing his Natural Selection to Artificial Selection in agriculture. He attempted to base his theory of Natural Selection on the achievement of artificial selection, which successfully produced new sorts of plants and domesticated animals. In Darwin's times, English farmers and breeders selected male and female specimens with the most desirable traits to breed offspring who would hopefully share and even improve those traits. Dozens of new breeds of pigs, sheep and cattle were produced that way.[105] Darwin presented Natural Selection as analogous to this, artificial one.

Natural Selection, however, differs from the artificial one in more than one way. Artificial selection is by design. As for Natural Selection, Darwin himself emphatically denied that it was a product of intelligent design: it had no author. There was no Logos detected behind it. On his part, Malthus had never believed that humans were creative or productive in their struggle. He also wrongly predicted the population's looming expiration from hunger. Despite our contemporary calamities and famines, we are now past Malthus' markers for our demise.

Both Malthus and Darwin discarded the power of science and reason. Malthus disregarded the role that reason and science played in enhancing human resources. Darwin also disregarded this role. He inferred the laws of Natural Selection by analogy with the artificial one, not the other way around. Usually, science learns the laws of nature and then applies them to technology. Darwin went the other way: he projected the artificial onto the natural. Yet the successful results of selection in stockbreeding, guided by science, do not guarantee the success of blind Natural selection. Artificial selection implies a method that Darwin emphatically denied to Nature. Tsvelik puts it this way: "Darwin took the most apparently teleological process available—the emergence of new forms of life,—and tried to prove that it can be reduced to purposeless process of Evolution by Random Mutations and Natural Selection. Thus, Darwin's analogy is too bold assuming that the process purposively organized by humans can go on by itself without any purpose." Referring to the recent study of Eugene Koonin, The Logic of Chance, Tsvelik thus criticized the logic of Darwin:

> Since there is an immense difference in complexity between living organisms and inert matter, life requires explanation. Darwin understood that

it could not be just chance, even in his times scientists had been sufficiently sophisticated to see an utter improbability of chance bridging of the gap between, say, a cat, and a stone. So, Darwin postulated that this gap has been bridged by an incremental process, by what Richard Dawkins calls "climbing Mount Improbable". That is what Natural Selection is. It maintains that although the chance of any beneficial mutation is small, the gain in reproductivity may be so large as to outweigh this. As a counterargument Koonin provides a simple calculation demonstrating that natural selection (random mutations + selection of features beneficial for survival) can work ONLY IN VERY LARGE populations. Since by a sheer law of big numbers most of the mutations are detrimental, in small populations everybody will die sooner than any beneficial mutation will occur. However, evolution has taken place, the evidence for the change is overwhelming. If not Natural Selection, then what?

As Tsvelik further argues the existence of Evolution suggests contingency, both the general laws of physics and particular circumstances have been prearranged, the dice of the evolutionary process has been loaded in favor of its progressive development.

Though Darwin himself was not conscious of this implication, Henri Bergson brought it to the surface. In his *Creative Evolution*, Bergson maps the difference between Instinct and Intellect, both contributing to the noetic matrix of the *élan vital*. This term, coined by Bergson, refers to a highly complex, heterogeneous, and organic process, the source of all the manifestations of life. This *élan* moves in three main

ALEXEI TSVELIK * SIX DAYS

venues—plants, animals, and humans; each group, a specific vessel for the consciousness of the *élan*.[106] Thus, unlike Darwin, Bergson sees a conscious force behind Nature, which is fundamentally different from all human cultural, technical, and creative activity. Bergson's vision can easily agree with the implications of Darwin's vision, though Darwin would hardly agree with this last statement.

This one-way continuity demonstrates that Darwin, though a positivist, nevertheless endowed his key concept of Natural Selection with a metaphysical meaning, albeit failing to show how, precisely, it would work. Without a precise description of its mechanism, the concept of Natural Selection had little scientific value. Today's genetic and genomic science has dismantled Darwin's two main concepts: Natural Selection and one single tree of life. Eugene Koonin, an eminent specialist in evolutionary biology, calls Darwin's version of Evolution, an "oversimplified myth."[107] In the introduction to his *Logic of Chance*, Koonin writes:

> The genomic revolution did more than simply allow credible reconstruction of the gene sets of ancestral life forms. Much more dramatically, it effectively overturned the central metaphor of evolutionary biology, ... the Tree of Life ... , by showing that evolutionary trajectories of individual genes are irreconcilably different.[108]

Commenting on Darwin's *The Origin of Species*, Koonin also points out that Darwin himself failed to produce any scientific evidence for one single ancestor, or for any mutations between species: "In a rather striking departure from the title of the book, all indisputable examples of evolution that Darwin presented involve the emergence of new varieties within a species, not new species."[109] "Our unfaltering admiration for Darwin notwithstanding, we must relegate the Victorian world-view (including its refurbished versions that flourished in the twentieth century) to the venerable museum halls where it belongs."[110]

According to Koonin, the genomic revolution has dis-covered a plurality of patterns and processes, as well as the *central role of contingency* in the evolution of life forms. The conclusive point of Koonin's *Logic of Chance* is rather striking: "viewed from the vantage point on the current state of Evolutionary Biology," he argues, one has to acknowledge "the overwhelming importance of chance in the emergence of life on Earth."[111] This chance, to boot, is miniscule. Koonin calculates the probability of the emergence of life, on our planet, or in the universe as a whole, as 10^{-1018}.[112]

At the end of the day, Koonin arrives at a critique, analogous to Kant's. He questions over-generalizing sci-entific worldviews. All science can actually do, Koonin maintains, is to develop models and assess their predic-tive power. "The scientific process does not tell us any-thing directly about the world; it tells us only about the compatibility of certain observations with the adopted models. All aspects of any worldview ('the picture of reality') can be considered *metaphysical implications* of the models and, as such, inconsequential."[113] "No model can claim to accurately represent 'reality,' which is unknowable in principle."[114]

Though, not all scholars of genetics share such an agnostic position. Yuri Petrovich Altukhov, a member of the Soviet Academy of Science, at his lectures and sem-inars in the department of biology of Leningrad State University, used to tell his students:

I arrived at the conclusion of the existence of the Creator also because the experiments of my collaborators and my own demonstrated that not only the origin of man, but even of the ordinary biological species cannot be accidental. Each species strictly preserves its uniqueness. Its basic characteristics are determined not by polymorphism—as if it is a petty cash which a species pays for getting adapted to its environment. No, the most important vital properties of a species are determined by the monomorphous part of the genome, which constitutes the basis of uniqueness of species: accidental changes of these genes are lethal. Therefore our world cannot be result of natural selection.[115]

Such is one contemporary critique of Darwin's theory. It is a myth in the whole series of modern myths. The French Enlightenment, for example, substituted mechanistic mythology for the belief in Divine creation. Darwin replaced the same belief with his own mythology—of Natural Selection.

Dialectic Materialism

Darwin's theory became a key to the Marxist doctrine of Dialectic Materialism. Engels used Darwin's hypothesis in his own theory of anthropogenesis. Though an intellectual himself, Marx found the main source of economic productivity not so much in the mind of scientists, like James Clerk Maxwell for example, but in workers' physical labor. Engels also ascribed the chief role in anthropogenesis to physical labor, including the hand as its human instrument. He explicated this theory in his *Dialectic of Nature*, in the chapter titled "The Part Played by Labor's Role in the Transition from Ape to Man." This myth, traceable to Engels, was the foundation of Soviet education. Moreover, it was developed and completed, neither by Marx nor Engels themselves, but by later Soviet Marxist ideologues.

As the term Dialectical Materialism suggests, it purports to synthesize materialism with dialectics. Marx had been an enthusiastic student of Hegel's thought before he converted to materialism under Feuerbach's influence. He, realized, however, that materialism could not account for the human agent or for the subjectivity of human cognition and practice.[116] It needed to be supplied with dialectics inasmuch as Hegel himself had acknowledged the movement of dialectics both in thought and in natural and historical processes. In Marx's view, Hegel had erroneously deified *Logos*, considering it the principal reality. Marx claimed to set Hegel back "on his feet" by rejecting such a metaphysical entity as *Logos*: dialectics does not need *Logos*, being inherently present in the process of reality as such. The human mind merely reflects dialectics while thinking properly about material reality of Nature and history.[117]

For Hegel, however, it is the *Logos* that is the driving force of dialectics. The human mind, *Nous*, naturally constrained by its own limitations, tends to stick to its own concepts, which are mental abstractions of reality. The *Logos*, in its all-encompassing presence, challenges the limitations of *Nous* through dialogical encounter with other mind's abstractions of reality, albeit also one-sided, thus driving the *Nous* forth to overcome its confines by digesting the opposite view(s). Thus *Logos*, by its invisible activity through the endless challenges of ever new dialogical encounters, makes *Nous* transcend itself in ever new syntheses. This is a thoroughly noetic process which, without *Logos*, would extinguish itself.

Marx, however, wanted to set Hegel "on his feet" only in his political-economic doctrine. It was Engels who attempted to develop a more pervasive dialectical system upon materialistic premises. At the outset of his enterprise, Engels tried to preserve the open-ended method of dialectics. Since the world of Nature and human actions consists of processes, and human knowledge reflects these processes, science itself is a process which never ends and cannot attain the absolute truth. Thus, Engels claims to offer not a closed doctrine, but a method of "dialectically-advancing progressive scientific knowledge of reality which is always open to further change and development."[118]

This did not happen for several reasons. Dialectics emerged, was recognized and developed as the activity of the human mind, and as the timeless inherent laws of human thinking. But Engels, like Marx, regarded dialectics as a purely material process and claimed to make it "simple and clear as noonday." He presents it as "nothing else but the science of the general laws of movement and development in Nature, human society and thought."[119] He raised a rather apt question: how can the universe "be arranged in accordance with a system of thought which itself is only the *product of a definite stage of evolution of human thought?* (Emphasize mine—MM)."[120]

When Engels set out to rework Hegel's principles in his *Dialectics of Nature*, however, these principles, legitimate in the realm of the Logos, become obscure and illogical in his metaphysics of matter. Dialectical materialism turns out to be an oxymoron. *Nous*, finding itself without the *Logos'* driving challenges, finds itself locked in closed dogma. This is exactly what happened with dialectical materialism. Engels failed to produce a consistent philosophical system, and, though he worked on *Dialectics of Nature* (*Dialektik der Natur*) for ten years (1872–82), he never finished it. After his death, other German socialist thinkers tried to publish his unfinished manuscript (s), but failed to find a publisher.

Marx and Engels did not complete the philosophical project they had initiated, nor did they christen it as *Dialectic Materialism*. The name emerged in German socialist circles and was applied to their philosophy after their death; Karl Kautsky, for example, used this expression in his biography of Engels. By that time, Dialectic Materialism was already becoming an ideological doctrine. It might have happened without Marx' and Engels's explicit intention, since they had wanted their materialist dialectics to remain open-ended. But designing their philosophy to be a practical political instrument for "changing the world" they prearranged its fate as ideological dogma in the hands of Communist Party.

And so it happened. Engels's unfinished manuscript (s) of *Dialectics of Nature* landed in the Soviet Institute of Marx and Engels, where it was doctored into a book published in 1927. This book itself became a cornerstone of the Soviet doctrine of Dialectical Materialism. With Lenin's and Stalin's contributions, Marx's and Engels's writings became the ideological basis of Communist regimes all around the world, beginning with the Soviet Union. The Soviets used this very doctrine to purge or oppress Science, including its studies of complex evolution in Nature, especially of genetics.

Although Soviet ideological history is now history, the myth of Dialectical Materialism still infiltrates our minds today. This myth, however, does not help us to understand how matter has evolved into life and human civilization. Not that Darwin or Engels themselves were happy about this ideological outcome. What ailed both of them was their own fear that the painful and impersonal evolutionary process resulting in the appearance of humanity was too random to protect life on earth from the doom of annihilation. The former Christian faith would provide some home, but they lost it.[121]

The Ring of Recurrence

Friedrich Nietzsche (1844–1900) further perpetuated this myth by accepting it as a scientifically proven reality. A Classical philologist and humanist, Nietzsche was horrified by the vision of an unequal struggle—between man, with his culture, on the one hand, and the blind, brutal force of Natural Selection, on the other. Nietzsche saw man as too delicate and complicated a creature to survive in this struggle. If so, man would have to be overcome, and renounce his cultural sophistication, returning to the primeval state of his animal struggle for survival. The "Death of God" that Nietzsche announced in his "Gay Science," entailed the disappearance of "Heaven" as Culture,—man's treasury of Platonic ideas and philosophers' Reason. Compared to the cold, endless Universe, the most sophisticated bloom of human civilization is no more than an accidental growth, doomed and aimless.

To alleviate his despair, Nietzsche came up with the proclamation of eternal recurrence. He believed in the scientific myth of his age, mostly in Spencer's version of it. Spencer blindly accepted what he trusted to be a foolproof scientific fact—that in the endless universe the same development would repeat itself at gigantic intervals. Nietzsche turned this hypothesis of "the ring of recurrence" into his own message of "salvation."[122]

Today, Nietzsche's eternal recurrence is revisited in the hypothesis of a multiverse—a multitude of universes beyond and besides our infinitely expanding universe. Nietzsche proposed the message of eternal recurrence as the destiny of the universe, i.e., a sort of teleology his own. Nietzsche, however, had a literary predecessor—a character from Dostoevsky's *Brothers Karamazov*." While going insane, Ivan Karamazov has a nightmare in which the Devil sarcastically caricatures this proto-Nietzschean teleology, carrying it *ad absurdum*:

> But our present earth may have repeated itself a billion times; it died out, let's say, got covered with ice, cracked, fell to pieces, broke down into its original components, again there were the waters above the firmament, then again a comet, again the sun, again the earth from the sun—all this development may already have been repeated an infinite number of times, and always in the same way, to the last detail. A most unspeakable bore...[123]

All of these worldviews, allegedly based on Science— be it Laplace's Mechanistic Universe, Spencer's Eternal Development, or Darwin's Biological Evolution, and the new myth of the multiverse—have one thing in common: the human person remains forlorn in the infinite space and time of a purposeless development. Both the human *Nous* and the Divine *Logos* are lost in this process without beginning or end. In the Appendix to his book, Tvelik discusses both the Anthropic Principle, and the theory of the Multiverse and Chaosegenesis. He considers it as a

ALEXEI TSVELIK * SIX DAYS

postmodern myth which constitutes a departure not just from all principles of Science, but from the principles of rationality as well. "Science, he maintains, "has never claimed that *everything is possible*, as the adherents of this myth do."

A Vindication of Science

Today, the Christian Church takes into consideration the scientific idea of Evolution understood as the gradual emergence of new forms. Relying on both Science and Biblical scholarship, the Church no longer considers the Biblical Hexameron as a rival to scientific discourse, just as it does not consider the days in Genesis 1 in terms of the astronomical 24-hour ones. For the Church, the belief in Divine Creation neither promotes nor contradicts Evolution. The two discourses belong to different genres. The main difference is that the Church sees the cosmological process as a purposeful one. The aim of Tsvelik's book is to demonstrate that one can infer this purpose from the scientific data.

Vladimir Solovyov, a contemporary of both Spencer and Darwin, considers Evolution in his philosophical teaching on Divine Humanity. Even beyond his Lectures dedicated to the topic, his doctrine of Divine Humanity implies an evolutionary process, inasmuch as it brings about the unity of the whole material world by permeating it with Reason. Every stage of Evolution provides the basis for the next one. It develops the organs and instruments necessary for the next and higher stage. Thus, according to Solovyov, inorganic matter constitutes the foundation for vegetable functions in life; these vegetable functions, in turn, provide the ground for animal functions for all, including humans; these animal functions, in turn, make human mental activity possible. Since each lower stage is incorporated into the next, higher one, Evolution "is not merely a process of development and expansion but also the process of the unification of the universe."[124] This unification is completed in man and by man as its agent and subject. Thus, Solovyov transforms Spencer's and Darwin's theory of blind evolution, while introducing the Anthropic Principle into the process.

The higher stages of life follow the lower ones only chronologically. Metaphysically, they precede them. In terms of Logos, the Higher Reason, every new stage of Evolution is a new act of creation, in and by this Logos.[125] As a reasoning creature, man is the apex of Evolution and therefore fit for Divine Incarnation. This Solovyovian ontology, of course, presupposes a non-linear understanding of time in the Bible.

Here Prof. Tsvelik concurs with Solovyov and makes the Anthropic Principle the central idea of his book. Tsvelik's explication allows for comparing and contrasting various myths for the same truth—of Creation as Evolution. Without devaluing any of these myths, based on his explication, we can nonetheless address their discourse on this one truth as precisely that—of various myths, be they Darwinian, Encyclopedist, or Dialectical Materialist. The Biblical Hexameron is neither better nor worse than any of those other myths or genres of discourse. It has, though, one advantage—it provides meaning for our existence. Therefore, its symbolism does merit consideration.

In Tsvelik's presentation, today's Science demonstrates that Cosmogony constitutes a sophisticated laboratory. This laboratory is designed and fashioned in a way that makes possible the emergence of man as a reasoning creature, capable of cognizing and understanding this laboratory itself. According to Tsvelik, Science reveals the universe as commensurate with man. Tsvelik's *Six Days* and the *Hexameron* of the Bible agree that *Homo sapiens* is created in the image and likeness of God and according to His Logos, a higher Reason, providing the blueprint for life.

Notes

1   This point was most forcefully made by P. K. Feyeranbend, "Science in a Free Society", London: New Leaf Books, 1978, p.70. See also Thomas Kuhn, the 1969 postscript to "The Structure of Scientific Revolutions", p. 198ff.

2   It is mathematically accurate to call this continuum one plus three dimensional since time dimension is not completely equivalent to space ones.

3   The relative temperature variations constituted 1/1000 percent.

4   See Appendix C for explanations.

5   "This leads us to consider a remarkable—indeed *fantastical*—thing about the Big Bang. It is not merely a mystery of its occurrence, but that it was an event of extraordinarily low entropy. Moreover, the remarkable thing is not merely that but the fact that the entropy was low in a very particular way, and apparently *only* in that way, namely that the *gravitational* degrees of freedom were, for some reason, *completely suppressed.* This is in stark contrast with the matter degrees of freedom and those of (electromagnetic) radiation, as they appear to have been maximally excited in the form of a thermal, maximal entropy state. To my mind, this is perhaps the most profound mystery of cosmology and, for some reason, is a still largely unappreciated mystery!" R. Penrose in *Fashion, Faith and Fantasy in the New Physics of the Universe,* ch.3

6   The entropic considerations have a direct bearing on the idea of the Beginning. Even if quantum effects allowed the universe to avoid a

singularity at the time of the Big Bang, as many physicists believe, entropic considerations do not support the idea of an eternal universe developing through an infinite series of big bangs.

7   The newly fashionable arguments of multiverse theory will be dealt with in Appendix A.

8   Cf., Denton 2017.

9   See, for instance, Martin Chaplin, "Anomalous properties of water" at http://www1.lsbu.ac.uk/water/water_anomalies.html. Accessed 4/3/2019.

10   "Unit 8. Climate: Atmosphere and Oceans" at http://resilience.earth.lsa.umich.edu/Inquiries/Inquiries_by_Unit/Unit_8.htm. Accessed 4/3/2019.

11   Clyde A. Hutchinson III et al., "Design and synthesis of a minimal bacterial genome," *Science* 351 (25 Mar 2016) at http://science.sciencemag.org/content/351/6280/aad6253.full.pdf+html. Accessed 4/3/2019

12   The interested reader will benefit from the review of current theories of the origin of life in the book by the prominent evolutionary microbiologist Eugene Koonin (2012).

13   Tia Ghose, "3.5-Billion-Year-Old Fossil Microbial Community Found," (November 13, 2013), http://www.livescience.com/41191-ancient-microbe-fossils-found.html. Accessed 4/4/2019

14   See, for example, Deborah Byrd, "Is it true that Jupiter protects Earth?" (November 25, 2015) https://earthsky.org/space/is-it-true-that-jupiter-protects-earth. Accessed 4/4/2019

15   Henri Atlan. "The Living Cell as a Paradigm for Complex Natural Systems." Complexus 1(1) (January 2003): 1–3. To get a better impression of this process, the reader is advised to watch the animation: https://www.wehi.edu.au/wehi-tv/dna-central-dogma-part-1-transcription

16   As quoted in "On 'computabilism' and physicalism: Some Problems" by Hao Wang, in *Nature's Imagination* (1995), edited by J. Cornwall, p.161-189

17   Several theories of evolution have emerged during the recent 30 years: the theory of self-organization, the theory of neutral evolution, the so-called evo-devo theory, the theory of epigenetic inheritance, natural genetic engineering, the theory of intellectual design.

18   To be fair this assumption is not that crucial and has been challenged already within the Darwinian paradigm by such evolutionary biologists as Steven J. Gould who argued that the speed of evolution can strongly vary.

19 See, for instance, the discussion in Chapter 1 of *The Logic of Chance: The Nature and Origin of Biological Evolution* (2012) by Eugene V. Koonin.

20 D. Dennett, *Darwin's Dangerous Idea*, p. 50.

21 There was a time when these "dark" portions which do not code for proteins had been considered as "junk", that is as remnants of unsuccessful mutations. The presence of these "junk DNA" was considered as an important evidence for the validity of STE. Later it was discovered that contrary to these early conclusions more than 80% of human genome perform important biological functions "dispatching the widely held view that the human genome is mostly 'junk DNA'". Ecker J. R., Bickmore W. A., Barroso I., Fritchard J. K. Gilad Y., and Segal E., "Genomics: ENCODE Explained", *Nature* 489, 52-55 (2012).

22 "The fossil record suggests that the major pulse of diversification of phyla occurs before that of classes, classes before that of orders, orders before that of families...The higher taxa do not seem to have diverged through an accumulation of lower taxa." Erwin D., Valentine J., and Sepkowski J. J. "A Comparative Study of Diversification Events: The Early Paleozoic versus the Mezozoic", Evolution 41, 1177-86 (1987).

23 Bowring S. A., Grotzinger J. P. Isachsen C. E., Knoll A. H., Pelechaty S. M., and Kolosov P., "Calibrating Rates of Early Cambrian Evolution" *Science* 261, 1293-98 (1993); Erwin D.H., Laflamme M., Tweedt S. M., Sperling E. A., Pisani D., and Peterson K. J. "The Cambrian Conundrum: Early Divergence and Later Ecological Success in the Early History of Animals", *Science* 334, 1091-97 (2011); see also McMenamin M. A. S. and McMenamin D. L. S., *The Emergence of Animals: The Cambrian Breakthrough*. New York: Columbia University Press, 1990.

24 As quoted in The Observer (11 January 1931); also in Psychic Research (1931), Vol. 25, p. 91

25 Thomas Nagel, *"Mind and Cosmos: Why the Materialist Neo-Darwinian Conception of Nature is Almost Certainly False" (2012)*.

26 As for Internal World (IW), see the discussion in the previous Chapter.

27 I just would like to emphasize that all operations related to space travel, be it manned or unmanned, require extreme precision and sophistication.

28 This thesis is supported by the declaration by Richard Lewontin, made in his review of Carl Sagan's "The Demon-Haunted World"

in the *New York Review of Books*, January 9, 1997: "It is not that methods and institution of science somehow compel us to accept a material explanation of the phenomenal world, but, on the contrary, that we are forced by our *a priori* adherence to material causes to create an apparatus of investigation and a set of concepts that produce material explanations, no matter how counter-intuitive, no matter how mystifying to the uninitiated. Moreover, that materialism is absolute, for we cannot allow a Divine Foot in the door".

29 Consider this: "The rise of modern science coincides with the suppression of non-Western tribes by Western invaders. The tribes are not just physically suppressed, they also lose their intellectual independence and are forced to adopt the bloodthirsty religion of brotherly love—Christianity... Today this development is reversed... But science still reigns supreme... Thus, while an American can now choose the religion he likes, he is still not permitted to demand that his children learn magic rather than science at school... And yet science has no greater authority than any other form of life." P. K. Feyerabend, "Against Method: Outlines of an Anarchist Theory of Knowledge", London: New Leaf Books, 1975. "A thought of any kind is grounded in society... The individual, then, derives his worldview socially in much the same way that he derives his roles and his identity. In other words, his emotions and his self-interpretation like his actions are predefined for him by society, and so is his cognitive approach to the universe that surrounds him". Peter Berger, "Invitation to Sociology", Garden City; Doubleday & Co., Inc. 1963, p. 117.

30 A. Einstein, *Letters to Solovine*, translated by Wade Baskin, with an introduction by Maurice Solovine (New York: Philosophical Library, 1987), pp. 132-133.

31 One could include in this picture the gravitational force exerted by the Earth, but this will not change the logic of the argument.

32 Roger Penrose discusses the topic of uniformity at length in *Fashion, Faith and Fantasy in the New Physics of the Universe* (2016). See also Appendix C.

33 Translated by L. Pevear. "Ты видишь, ход веков подобен притче и может загореться на ходу".

34 See, for example, the book by Frank Wilczek, *A Beautiful Question: Finding Nature's Deep Design* (2015).

35 I will allow myself an extended quote from an essay by Alexei and Lev Burov [11]: "First, the laws are endowed with a peculiar mathematical beauty, uniting in themselves formal simplicity, rich-

ness of solutions and one or another kind of symmetry, often as if suggesting itself as a hypothesis to a mind gifted with intuition. This special beauty is sometimes called elegance of the laws of nature. Thus, elegance has a decisive significance to a birth of a hypothesis, the most mysterious part of discovery. Secondly, the same elegant mathematical law captures a tremendous range of parameters (distances, energies, etc.), at that with a fantastic precision, up to twelve digits. This quality of the laws can be called universality. Finally, the laws happen to be friendly to life's appearing and developing up to intellect; following the established terminology, this quality can be called anthropic. The combined presence of these three qualities allowed for their discovery by great minds, and for that reason, it seems that the most appropriate term, uniting all three, is discoverability. A universe whose laws satisfy the *Discoverability Principle* of being elegant, universal and anthropic we suggested to call *Pythagorean*. It could be even that the laws of our universe constitute the simplest possible set, compatible with the Discoverability Principle. The only so far available explanation of this amazing quality of the laws is that they come from the highest mind that created our universe able to not only be inhabited by intelligent beings but cosmically cognized by them (2016-2017, 8–9)." Discoverability requires the laws to be complicated enough to support possibilities for embodied reasonable beings and, at the same time, to be simple and elegant enough to be discoverable by these very beings (private communication).

36 I refer the reader to the brilliant book by Mark Steiner, *The Applicability of Mathematics as a Philosophical Problem* (1998) for more details.

37 They determine, for instance, the fact that the force between two electric charges in three-dimensional space is inversely proportional to the square of the distance.

38 Mark Tegmark, "The Mathematical Universe", https://arxiv.org/pdf/0704.0646.pdf

39 The reader interested in the technical details may look at the discussion on http://physics.princeton.edu/~cosmo/sciam/index.html#faq

40 The reader may find similar arguments put forward in the essay *Existing Because Ethically Required* by philosopher John Leslie (in *The Mystery of Existence: Why There is Anything at All*, 2013).

41 The attentive reader might have noticed that in the present context I do not distinguish between equations and inequalities. I believe

that this distinction becomes irrelevant when the number of conditions is almost infinite, as it is.

42   I am infinitely grateful to M. Arkadiev, who clarified for me that aspect of the Anthropic Principle.

43   Here is the whole quote: "Even if there is only one possible unified theory, it is just a set of rules and equations. What is it that breathes fire into the equations and makes a universe for them to describe? The usual approach of science of constructing a mathematical model cannot answer the questions of why there should be a universe for the model to describe. Why does the universe go to all the bother of existing?"

44   It is his own paraphrase from "Genesis of a Pythagorean Universe".

45   The number of neutrons in the nuclei of a given element can vary; nuclei with different numbers of neutrons are called isotopes. For example, we know of three isotopes of hydrogen: ordinary hydrogen ^{1}H, whose nuclei consist of just one proton and no neutrons, deuterium ^{2}De, whose nuclei consist of one proton and one neutron, and tritium ^{3}T, whose nuclei consist of one proton and two neutrons. Usually only one isotope is stable and all the other isotopes decay with various half-lives and hence are less abundant.

46   Popper, Karl. 1978. "Natural Selection and the Emergence of Mind." Dialectica 32, No. 3/4: 339–55. http://www.information-philosopher.com/solutions/philosophers/popper/natural_selection_and_the_emergence_of_mind.html

47   This point is based on the research and formulation by Prof. Olga Meerson.

48   Cf., "Who is wise and understanding among you?" (James, 3:13), "If any of you lacks wisdom, let him ask God, who gives to all men generously and without reproaching, and it will be given him" (James, 1:5).

49   Cf., Gerhard von Rad, *Old Testament Theology*, New York: Harper & Row, 1962, vol. I, 418-9, 423, 428, 421.

50   St. Athanasius, On the Incarnation. In Adversus Gentes, Libri duo. §41:5; §42:7. *Nicene and Post-Nicene Fathers of the Christian Church*, Michigan, Grand Rapids: WM. B. Eerdmans Pub. Co. v. IV, 58-59.

51   Cf., *A History of the University in Europe*, in 4 volumes, Walter Ruegg, General editor, Cambridge University Press, 1996-2011.

52   Cf., V. F. Asmus' point in his *Immanuel Kant*, Academy of Science, SSSR, Moscow: Nauka, 1973, 117.

53   I. S. Dmitriev, *The Unknown Newton* (*Неизвестный Ньютон*), St. Petersburg: "Aleteiia," 1999, 17.

54   Maybee, Julie E., "Hegel's Dialectics", *The Stanford Encyclopedia of Philosophy* (Winter 2016 Edition), Edward N. Zalta (ed.) https://plato.stanford.edu/archives/win2016/entries/hegel-dialectics

55   Immanuel Kant, *Critique of Pure Reason*, trans. Paul Guyer and Allen W. Wood, Cambridge: Cambridge University Press, 1998, 387.

56   Asmus, op. cit.,186-7.

57   Cf., Frederick Copleston, S.J. *A History of Philosophy*, Image Books, Doubleday, 1985, VI, 189.

58   Cf., Asmus, op. cit., 85.

59   Cf., *Personalism*, Boston, 1908, by Borden Parker Bowne (1847–1910), a founder of the American tradition of philosophical personalism.

60   Fichte, Basis of the Entire Theory of Science, Part I, §1, 1-2; Cf., Copleston, A History of Philosophy, 7:49-50.

61   Copleston, 7:40.

62   Edmund Husserl, *Ideen zu einer reinen Phänomenologie und phänomenologischen Philosophie. Erstes Buch: Allgemeine Einführung in die reine Phänomenologie* (*Ideas: General Introduction to Pure Phenomenology*), 156. (par.57), quoted from Gustav Shpet, "Soznanie i ego sobstvennik" (Consciousness and its owner) in "*Filosofskie etudy*" (*Philosophical etudes*), Moscow: Progress, 1994, 88.

63   Edmund Husserl, *General Introduction to Pure Phenomenology, Routledge reprint of Taylor & Francis, 2002, p.156.*

64   Gustav Schpet, op. cit., 54.

65   Cf., John Searle, *The Rediscovery of the Mind*, Cambridge, MA: MIT Press, (1992), *passim*.

66   This point, though, through a different train of reflection, was also made by Fr. Pavel Florensky. Identifying Reason with Truth, he shows that the Truth cannot be just an impersonal entity, the Truth must be "*a self-proving Subject, a Subject qui per se ipsum concipitur et demonstratur* (that is conceived and proved through itself), a Subject that is absolute Lord of itself, that is master over the infinite series of all of its grounds, which are synthetized into a unity and even into a unit... The conception of God as having His being and reason in Himself runs through scholastic philosophy like a scarlet thread and finds its extreme but one-sided application in Spinoza..." Pavel Florensky, *The Pillar and Ground of*

the Truth, trans. Boris Jakim, Princeton, NJ: Princeton University Press, 1997, pp.34-35. This point is demonstrated step by step in the third chapter, "Letter Two: Doubt", 14-37.

67 John Macmurray, *The Self as Agent,* New York: Humanity Books, 1999, p.86.

68 Cf., V.I. Vernadsky, *Kant I estestvoznanie XVIII stoletia, (Kant and the Natural Science of the XVIII Century),* Moscow, 1905.

69 *Allgemeine Naturgesehichte und Theorie des Himmels,* (Russian transl. "Vseobshchaiia estestvennaia istoriia I teoriia neba," I. Kant, SS (Works), Moscow, ANSSR, Mysl', 1964, vol. I.

70 Cf., Asmus, op. cit., 12-14, 17-8.

71 Cf., A. Arseniev, A. Gulyga, "Rannie Raboty Kanta" (Kant's early writings), I. Kant, SS (Works), Moscow, 1964, vol. II, 15, 20-22.

72 Cf., Asmus, op. cit., 18, 24.

73 F. Copleston, History of Philosophy, VII, 172-173.

74 "The movement of the solar system takes place according to unchangeable laws. These laws are Reason, implicit in the phenomena in question. But neither the sun nor the planets, which revolve around it according to these laws, can be said to have any consciousness of them." Hegel's *Philosophy of History,* III. Philosophic History, i. Reason Governs the World, § 15: https://www.marxists.org/reference/archive/hegel/works/hi/history3.htm#i

75 Hegel, Encyclopaedia Logic (EL, Addition 1 to #24), quoted from Julie E. Maybee, "Hegel's Dialectics", *The Stanford Encyclopedia of Philosophy* (Winter 2016 Edition), Edward N. Zalta (ed.) https://plato.stanford.edu/archives/win2016/entries/hegel-dialectics

76 Plato, Phaedrus, 249c. Plato, *Collected Dialogues,* eds. Edith Hamilton & Huntington Cairns, Bollingen series LXXI, Princeton University Press, 1989, 496.

77 Phaedrus, 229c, 230d, in *Collected Dialogues,* 478,479.

78 Phaedrus, 265d, in *Collected Dialogues,* 511.

79 Cited from Maybee, Julie E. "Hegel's Dialectics."

80 "This unity differs from Kant's antinomies where their truth obtains in the tension between the opposites. Unlike Hegel, Kant discusses a static truth."—commentary by Olga Meerson.

81 Cf., Maybee, p.2.

82 Cf., Julie E. Maybee, "Hegel's Dialectics."

83 Cf., Copleston, 7: 180.

84 Cf., Schelling, "A Philosophy of Revelation."

85   Hermann Cohen, *Kants Theorie der Erfahrung (Kant's theory of experience)*, 2nd ed. Berlin, 1885. From Sergei Bulgakov, *Philosophy of Economy*, trans. Catherine Evtuhov, Yale University Press, 2000, 164.
86   Cf., Sergei Bulgakov, *Philosophy of Economy, 158-160.*
87   Ibid., 161-2.
88   Ibid.
89   Ibid.
90   Ernst Mach, "Science", 463-4; Quoted by Pavel Florensky in "Vodorazdely mysli" (Watersheds of thought), Works, Moscow: Pravda, 1990, vol. II, 109.
91   Y. M. Lotman, *"Fenomen iskusstva" (Phenomenon of Art) in "Kul'tura I vremy" (Culture and Time),* 153.
92   Bulgakov, Philosophy of Economy, Trans. C. Evtuhov, 83.
93   Cf., Schelling, *System of the Whole of Philosophy and of Naturphilosophie in Particular.*
94   "The case of Herbert Spencer's system is much to the point here. Rationalists feel his fearful array of insufficiencies. His dry schoolmaster temperament, the hurdy-gurdy monotony of him, his preference for cheap makeshifts in argument, his lack of education even in mechanical principles, and in general the vagueness of all his fundamental ideas, his whole system wooden, as if knocked together out of cracked hemlock boards—and yet the half of England wants to bury him in Westminster Abbey. Why? ... The noise of facts resounds through all his chapters, the citations of fact never cease, he emphasizes facts… It means the right *kind* of thing for the empiricist mind." William James, *Pragmatism and Other Writings*, Penguin Books, 2000, p.22-23.
95   Herbert Spencer, *First Principles of a New System of Philosophy (1862)*, 6th ed. p. 119, quoted in Copleston, 8:124.
96   Herbert Spencer, First Principles, quoted in Copleston, 8: 128.
97   Ibid, from the Russian translation, *"Osnovnye nachala,"* St. Petersburg, 1897, 331.
98   This is Tsvelik's commentary to my afterword.
99   Cf., Herbert Spencer, First Principles, pp. 149-51, 167, 175. See also Copleston, 8: 126-127.
100 Darwin, C. R. *The Autobiography of Charles Darwin 1809-1882. With the original omissions restored and with appendix and notes by his grand-daughter Nora Barlow.* London: Collins, 1958, p. 49. Cf., also Charles Darwin, *The Life of Erasmus Darwin*, Cambridge, UK, Cambridge University Press, 2003.

101 Спенсер, 'Progress: Its Law and Cause' published in Chapman's in 1857), on the basis of which he wrote his *First Principles of a New System of Philosophy* (1862).

102 Charles Darwin, Autobiography,New York: The Norton Library, 1969, 108-109.

103 Charles Darwin, *Idem*.

104 Ibid., 119-120.

105 Cf., "Darwinism", G. Platonov, *Encyclopedia of Philosophy*, Moscow: Gosudarstvennoie nauchnoe izdatel'strvo, 1960, v. I. 430-431.

106 Cf., Henry Bergson, *Creative Evolution,* London, 1954, pp. 115-17; 142-46.

107 Eugene Koonin, *The Logic of Chance: The Nature and Origin of Biological Evolution*, New Jersey: FT Press Science, 2012, 425.

108 Ibid, viii.

109 Ibid, 5.

110 Ibid., viii.

111 Ibid., xii.

112 Ibid., 435.

113 Ibid, 426.

114 Ibid., 428.

115 This is the testimony of Larissa Volokhonsky-Pevear, his former graduate student back in the early 1970-es. Such statement of a Soviet scientist could not be printed in the Soviet Union, and therefore we have only this testimony of academician Altukhov's views.

116 This was the first thesis of his criticism of Feuerbach. Cf., Karl Marx, "Theses on Feuerbach," in Karl Marx, *Selected Writings*, ed. David McLellan, Oxford University Press, 1977, p. 156.

117 Cf., Marx, Poverty of Philosophy, 1847, quoted in Copleston, ibid., 311.

118 Cf., Copleston, Op. Cit., 320.

119 Friedrich Engels, "Anti-Duhring," p. 144, from Copleston, Ibid., 320.

120 Frederick Engels, *Dialectics of Nature:* https://www.marxists.org/archive/marx/works/1883/don/ch01.htm

121 Cf., "Believing as I do that man in the distant future will be a far more perfect creature than he now is, is an intolerable thought that he and all other sentient beings are doomed to complete annihilation after such long-continued slow progress. **To those who fully admit the immortality of the human soul, the destruction of our world will not appear so dreadful**." (Emphasis mine, MM).

Charles Darwin, *Autobiography... ,* 92.
"All that comes into being deserves to perish. Millions of years may elapse, hundreds of thousands of generations be born and die, but inexorably the time will come when the declining warmth of the sun will no longer suffice to melt the ice thrusting itself forward from the poles; when the human race, crowding more and more about the equator, will finally no longer find even there enough heat for life; when gradually even the last trace of organic life will vanish; and the earth, an extinct frozen globe like the moon, will circle in deepest darkness and in an ever narrower orbit about the equally extinct sun, and at last fall into it. Other planets will have preceded it, others will follow it; instead of the bright, warm solar system with its harmonious arrangement of members, only a cold, dead sphere will still pursue its lonely path through universal space. And what will happen to our solar system will happen sooner or later to all the other systems of our island universe; it will happen to all the other innumerable island universes, even to those the light of which will never reach the earth while there is a living human eye to receive it. And when such a solar system has completed its life history and succumbs to the fate of all that is finite, death, what then? We do not know. **But here either we must have recourse to a creator**, or we are forced to the conclusion that the incandescent raw material for the solar system of our universe was produced in a natural way by transformation of motion which are by nature inherent in moving matter, and the conditions of which therefore also must be produced by matter, even if only after millions and millions of years and more or less by chance but with the necessity that is also inherent in chance." (Emphasis mine, MM), Engels, *The Dialectics of Nature.*
122 See "The Yes and Amen Song" of Zarathustra as his psalm to the eternal recurrence. In *Thus Spoke Zarathustra,* in *The Portable Nietzsche,* trans. Walter Kaufmann, Penguin Books, 1976, 340-3.
123 Fyodor Dostoevsky, *The Brothers Karamazov,* transl. Richard Pevear and Larissa Volokhonsky, New York: Vantage /Books, 1991, 644.
124 Vladimir Solovyov, *Opravdanie dobra (The Justification of the Good),* SS, 8: 211-2, 219-20.
125 Cf., Ibid, 218-9.

Alexei Tsvelik

Born 1954, in Samara, Russia. Theoretical physicist, PhD 1980, Fellow of American Physical Society (2002), a recipient of Alexander von Humboldt prize (2014), he was employed by Landau Institute for Theoretical Physics (Moscow), University of Oxford (1992-2001), since 2001 in Brookhaven National Laboratory. Published more than 200 papers in peer refereed journals including Science, Nature Communications, Proceedings of National Academy of Sciences and Physical Review Letters, two books in quantum field theory published by Cambridge University press.

Alexey Burov

After graduating from the Faculty of Physics of the Novosibirsk University in 1980, he worked at the Budker Institute of Nuclear Physics, where he defended his PhD thesis. Since 1997, Dr. Burov was working as an employee of Fermilab (USA). In 2011-2013, he was at CERN, with a two-year visit. Author of many journal publications on the physics of charged particle beams. Fellow of the American Physical Society. Organizer and host of the Fermilab Philosophical Society. Author of many philosophical publications in the journal "Friendship of Peoples" (Russia), a number of publications in the journal "Knowledge is Power" (Russia) and in the Theological Herald of the St. Petersburg Theological Academy. Prize winner of the Foundational Questions Institute (FQXi) for the philosophical article "Genesis of a Pythagorean Universe", written in collaboration with his son Lev. The author of the blog on snob.ru.

Michael Meerson

Born 1944 in Moscow, USSR. Education: MA, Department of History at Moscow State University, 1968; MD, St. Vladimir Theological Seminary, NY, 1984; PhD in Theology, Fordham University NY. The pastor of Christ the Savior Orthodox Church in NYC since 1978 to the present. Author of "The Trinity of Love in Modern Russian Theology", Franciscan Press 1998, and other books in English and Russian; as well as various articles in history, philosophy and theology, published in English, Russian, French, Italian and Spanish.

www.ingramcontent.com/pod-product-compliance
Lightning Source LLC
Chambersburg PA
CBHW050122280326
41933CB00010B/1199